# Yang-Mills Measure on Compact Surfaces

of the
American Mathematical Society

Number 790

# Yang-Mills Measure on Compact Surfaces

Thierry Lévy

November 2003 • Volume 166 • Number 790 (end of volume) • ISSN 0065-9266

**American Mathematical Society**
Providence, Rhode Island

2000 *Mathematics Subject Classification.* Primary 58D20, 81T13, 81T27, 60G60, 60F20.

---

**Library of Congress Cataloging-in-Publication Data**

Lévy, Thierry, 1976–
 Yang-Mills measure on compact surfaces / Thierry Lévy.
  p. cm. — (Memoirs of the American Mathematical Society, ISSN 0065-9266 ; no. 790)
 "November 2003, volume 166, number 790 (end of 3 numbers)."
 Includes bibliographical references.
 ISBN 0-8218-3429-0 (alk. paper)
 1. Yang-Mills theory. 2. Quantum field theory. I. Title. II. Series.
QA3.A57 no. 790
[QC174.52.Y37]
510 s—dc22
[530.14′35]                                                                                                                        2003057765

---

# Memoirs of the American Mathematical Society

This journal is devoted entirely to research in pure and applied mathematics.

**Subscription information.** The 2003 subscription begins with volume 161 and consists of six mailings, each containing one or more numbers. Subscription prices for 2003 are $555 list, $444 institutional member. A late charge of 10% of the subscription price will be imposed on orders received from nonmembers after January 1 of the subscription year. Subscribers outside the United States and India must pay a postage surcharge of $31; subscribers in India must pay a postage surcharge of $43. Expedited delivery to destinations in North America $35; elsewhere $130. Each number may be ordered separately; *please specify number* when ordering an individual number. For prices and titles of recently released numbers, see the New Publications sections of the *Notices of the American Mathematical Society*.

**Back number information.** For back issues see the *AMS Catalog of Publications*.

Subscriptions and orders should be addressed to the American Mathematical Society, P. O. Box 845904, Boston, MA 02284-5904, USA. *All orders must be accompanied by payment.* Other correspondence should be addressed to 201 Charles Street, Providence, RI 02904-2294, USA.

**Copying and reprinting.** Individual readers of this publication, and nonprofit libraries acting for them, are permitted to make fair use of the material, such as to copy a chapter for use in teaching or research. Permission is granted to quote brief passages from this publication in reviews, provided the customary acknowledgment of the source is given.

Republication, systematic copying, or multiple reproduction of any material in this publication is permitted only under license from the American Mathematical Society. Requests for such permission should be addressed to the Acquisitions Department, American Mathematical Society, 201 Charles Street, Providence, Rhode Island 02904-2294, USA. Requests can also be made by e-mail to reprint-permission@ams.org.

---

*Memoirs of the American Mathematical Society* is published bimonthly (each volume consisting usually of more than one number) by the American Mathematical Society at 201 Charles Street, Providence, RI 02904-2294, USA. Periodicals postage paid at Providence, RI. Postmaster: Send address changes to Memoirs, American Mathematical Society, 201 Charles Street, Providence, RI 02904-2294, USA.

© 2003 by the American Mathematical Society. All rights reserved.
This publication is indexed in *Science Citation Index*®, *SciSearch*®, *Research Alert*®, *CompuMath Citation Index*®, *Current Contents*®/*Physical, Chemical & Earth Sciences*.
Printed in the United States of America.

∞ The paper used in this book is acid-free and falls within the guidelines established to ensure permanence and durability.
Visit the AMS home page at http://www.ams.org/

10 9 8 7 6 5 4 3 2 1    08 07 06 05 04 03

# Contents

| | |
|---|---|
| Introduction | vii |
| **Chapter 1. Discrete Yang-Mills measure** | 1 |
| 1.1. Notation | 1 |
| 1.2. Discretization of a surface | 3 |
| 1.3. Discrete holonomy and gauge transformations | 5 |
| 1.4. Discrete Yang-Mills measure | 7 |
| 1.5. Conditional Yang-Mills measure | 8 |
| 1.6. Invariance under subdivision | 13 |
| 1.7. Invariance under area-preserving diffeomorphisms | 16 |
| 1.8. Two examples | 17 |
| 1.9. Discrete Abelian theory | 20 |
| **Chapter 2. Continuous Yang-Mills measure** | 35 |
| 2.1. Projective systems of probability spaces | 35 |
| 2.2. A Riemannian metric | 37 |
| 2.3. Piecewise geodesic Yang-Mills measure | 39 |
| 2.4. Lassos and small piecewise geodesic loops | 41 |
| 2.5. The space of paths | 45 |
| 2.6. Definition of the random holonomy | 51 |
| 2.7. Consistency with the discrete theory | 57 |
| 2.8. Binding up the conditional versions | 61 |
| 2.9. Surfaces with boundary | 62 |
| 2.10. The random holonomy process | 65 |
| **Chapter 3. Abelian gauge theory** | 75 |
| 3.1. The random holonomy as a white noise functional | 75 |
| 3.2. Small scale structure of the Yang-Mills field | 79 |
| **Chapter 4. Small scale structure in the semi-simple case** | 85 |
| 4.1. Asymptotic independence: a zero-one law | 86 |
| 4.2. Asymptotic independence on the plane | 94 |
| **Chapter 5. Surgery of the Yang-Mills measure** | 99 |
| 5.1. Markov property of the Yang-Mills field | 99 |
| 5.2. Sewing two surfaces | 102 |
| 5.3. Mending a surface | 109 |
| 5.4. An area-dependent topological field theory | 115 |
| Bibliography | 121 |

# Abstract

In this memoir[1] we present a new construction and new properties of the Yang-Mills measure in two dimensions.

This measure was first introduced for the needs of quantum field theory and can be described informally as a probability measure on the space of connections modulo gauge transformations on a principal bundle. We consider the case of a bundle over a compact orientable surface.

Our construction is based on the discrete Yang-Mills theory of which we give a full acount. We are able to take its continuum limit and to define a pathwise multiplicative process of random holonomy indexed by the class of piecewise embedded loops.

We study in detail the links between this process and a white noise and prove a result of asymptotic independence in the case of a semi-simple structure group. We also investigate global Markovian properties of the measure related to the surgery of surfaces.

*Key words and phrases.* Gauge theory, Yang-Mills, continuous limit, multiplicative random holonomy, white noise, Markov random field, zero-one law.

---

[1] Received by the editor January 15, 2001, and in revised form December 4, 2002.

# Introduction

In 1954, Chen N. Yang and Robert L. Mills proposed to extend the formalism of quantum electrodynamics to the treatment of isotopic spin and this initiated the study of non-Abelian gauge theories. In [45], one finds in particular the Lagrangian density for the action which now bears their names. This action is a gauge-invariant non-negative functional

$$S : \mathcal{A} \longrightarrow \mathbb{R}_+ \tag{1}$$

defined on the space of gauge fields, that is, the space of connections on a certain principal $SU(2)$-bundle. For a given connection $A$, $S(A)$ is essentially the $L^2$ norm of the curvature of $A$. R. Feynman's path integral quantization procedure applied to this gauge theory requires the definition of integrals of the form

$$\int_{\mathcal{A}} H(A) e^{\frac{i}{\lambda} S(A)} \, dA, \tag{2}$$

where $\lambda$ is a parameter, $H$ some gauge-invariant functional of a connection and $dA$ is the "Lebesgue" measure on $\mathcal{A}$.

The subject of the present work is the study of the integral (2) in the Euclidean setting, that is, when $\lambda$ is purely imaginary, say $\lambda = -2i$. One speaks in this case of stochastic quantization of the gauge theory, as one seeks in fact to define integrals against the probability measure

$$d\mu(A) = \frac{1}{Z} e^{-\frac{1}{2} S(A)} \, dA, \tag{3}$$

where $Z$ is the normalization constant.

This expression does unfortunately not make much sense, except for the exponential term that we are going to explain in detail in the first part of this introduction. In a second part we will describe the class of functionals one expects to integrate against the Yang-Mills measure. Then we shall review a construction of the measure due in its final form to A. Sengupta [38] and finish by presenting briefly the content of each chapter.

### Basics of gauge theory

We consider a principal $G$-bundle $P \longrightarrow M$, where $M$ is a compact smooth manifold and $G$ a compact Lie group. The frame bundle of an orientable Riemannian manifold is a good example of an $SO(n)$-bundle, where $n$ is the dimension of $M$. A connection on $P$ can be defined as a differential 1-form denoted by $\omega$ on the total space $P$ of the bundle, taking its values in the Lie algebra $\mathfrak{g}$ of $G$ and satisfying a set of affine conditions (see [30] for a complete definition).

If one wants to work on $M$, as physicists customarily do, one must pull $\omega$ back locally to open subsets of $M$ by means of local sections of $P$, that is, local smooth choices of a reference frame. This is called *gauge fixing*. Let us choose such a section $s$ on an open subset $U$ of $M$. The 1-form $A = s^*\omega$ is defined on $U$ and depends on the choice of $s$. In the case of the frame bundle, $A(X)$ is the infinitesimal rotation corresponding to an infinitesimal parallel transport in the direction of the vector $X$, measured with respect to the reference frame. Changing this reference frame amounts to choosing a $G$-valued function $g$ on $U$ and replacing at each point $m$ of $U$ the frame $s(m)$ by its image under $g^{-1}(m)$, denoted[2] by $(s \cdot g)(m) = g^{-1}(m)s(m)$. This transforms $A$ into $A^g = (s \cdot g)^*\omega$ according to the *gauge transformation* formula

$$A^g = g^{-1}Ag + g^{-1}dg. \tag{4}$$

It may be useful to clarify what is meant by a gauge transformation, for this can refer to two closely related, but distinct, things. We have just described the first one: it is the change of the reference frame through which one looks at the objects defined on $P$. But there exists on $P$ a group of diffeomorphisms called the gauge group (see footnote 3 page 3 for a description of the gauge group in a simple case), denoted by $\mathcal{J}$, which acts by pull-back on the differential objects defined on $P$, in particular on the connections. If $\phi$ is an element of $\mathcal{J}$ and $\omega$ is a connection on $P$, then one can compare $\omega$ and $\phi^*\omega$ through the same local section $s$ of $P$, that is, compare $s^*\omega$ and $s^*(\phi^*\omega)$. However, this last form is nothing but $(\phi \circ s)^*\omega$, which is equal by definition to $A^g$, provided one defines $g$ by the relation

$$\phi \circ s = s \cdot g. \tag{5}$$

This shows that the two points of view are equivalent. Hence, by a gauge-invariant object, a functional for example, one means either a functional of the pull-back $A = s^*\omega$ of the connection which does not depend on the choice of $s$ or a functional on $\mathcal{A}$ which is invariant under the action of $\mathcal{J}$.

The curvature of a connection is a $\mathfrak{g}$-valued differential 2-form on $P$, defined by $\Omega(X,Y) = d\omega(X,Y) + [\omega(X), \omega(Y)]$ and its pull-back by a section $s$ is usually denoted by $F$. It is related to $A$ through

$$F(X,Y) = dA(X,Y) + [A(X), A(Y)]. \tag{6}$$

The transformation rule for $F$ is simpler than that of $A$: if $F^g$ denotes $(s \cdot g)^*\Omega$, then one has

$$F^g = g^{-1}Fg. \tag{7}$$

The point in (7) is that the norm of $F(X,Y)$ is well-defined, provided $\mathfrak{g}$ is endowed with a scalar product invariant by adjunction. Such a scalar product always exists under the assumption that $G$ is compact. For example, if $G = SO(n)$, then $\mathfrak{g}$ is the space of antisymmetric $n \times n$ matrices with trace equal to zero, and the Killing form $K(A,B) = -\operatorname{tr}(A^tB)$ is a possible choice.

Now if $M$ is an oriented Riemannian manifold of dimension $n$, then $F \wedge *F$ is a locally defined $n$-form with values in $\mathfrak{g} \otimes \mathfrak{g}$, which can be contracted using the scalar product, denoted by $\langle \cdot, \cdot \rangle$, into a globally defined real-valued $n$-form $\langle F \wedge *F \rangle$. The Yang-Mills action can be defined as

$$S(\omega) = \int_M \langle F \wedge *F \rangle. \tag{8}$$

---

[2]The action of $G$ on $P$ is denoted on the right.

Although $S$ is perfectly well-defined on $\mathcal{A}$, (3) does not make much sense: on the infinite-dimensional affine space $\mathcal{A}$, there is no translation-invariant measure[3] $dA$. Moreover, the invariance of the scalar product on $\mathfrak{g}$ and the transformation rule (7) imply the gauge-invariance of $S$. Thus, $Z$ should be proportional to the volume of $\mathcal{J}$, hence infinite.

This suggests a slight modification of the point of view, which is in fact very natural from a physical point of view: we replace $\mathcal{A}$ by its quotient under the action of $\mathcal{G}$. This restricts the scope to gauge-invariant functionals, the only ones to be physically meaningful. Let us take a closer look at these functionals.

## Gauge-invariant functionals

A large and natural family of gauge-invariant functionals is provided by the *Wilson loops*. If $l : [0,1] \longrightarrow M$ is a piecewise $C^1$ loop, a connection $\omega$ allows to define a parallel transport or *holonomy* along $l$. If $A$ is defined in a neighbourhood of $l$ as a pull-back of $\omega$, then this parallel transport can be represented by the element $\mathrm{hol}(A, l)$ of $G$ defined as $h_1$, where $h : [0, 1] \longrightarrow G$ is the solution to the differential equation

$$\dot{h}_t h_t^{-1} = -A(\dot{l}_t) , \quad h_0 = 1. \tag{9}$$

Sometimes, one writes

$$\mathrm{hol}(A, l) = P \exp \int_l A, \tag{10}$$

and $P$ is called the path-ordering operator. This is purely formal insofar $\mathrm{hol}(A, l)$ is not a function of $\exp \int_l A$, unless $G$ is Abelian.

If one chooses another local representation $A^g$ of the connection $\omega$, then one gets another element of $G$ related to the first one by

$$\mathrm{hol}(A^g, l) = g(l_0)^{-1} \mathrm{hol}(A, l) g(l_0). \tag{11}$$

The gauge-invariant quantity is thus the *conjugacy class* of $\mathrm{hol}(A, l)$: we denote it by $\mathrm{hol}(\omega, l)$. If $\alpha : G \longrightarrow GL_n(V)$ is a linear representation of $G$, then the quantity

$$W_{\alpha, l}(\omega) = \chi_\alpha \mathrm{hol}(\omega, l) \tag{12}$$

is well-defined, with $\chi_\alpha(g) = \mathrm{tr}\, \alpha(g)$ and $W_{\alpha, l}$ defines a complex-valued functional on $\mathcal{A}/\mathcal{J}$ which is called the Wilson loop associated with $\alpha$ and $l$.

It is usually assumed in the physical litterature that $\{W_{\alpha, l}\}$ is a complete set of observables, in that these functions separate the points on $\mathcal{A}/\mathcal{J}$. Two results proved by A. Sengupta in [36, 37] will allow us to discuss this statement.

We need first to define a finer gauge-invariant functional than the conjugacy class of the holonomy along a loop. To see how this can be done, observe that the holonomies along two loops based at the same point are conjugated by the same element of $G$ when one replaces $A$ by $A^g$. Given $n$ loops $l_1, \ldots, l_n$ based at the same point, this leads us to define $\mathrm{hol}(\omega, l_1, \ldots, l_n)$ as an orbit of the adjoint action of $G$ on $G^n$ defined by

$$\mathrm{Ad}(g)(g_1, \ldots, g_n) = (gg_1g^{-1}, \ldots, gg_ng^{-1}). \tag{13}$$

---

[3] One can however make sense of $dA$, very much in the same spirit as one usually interprets the measure $\mu$, namely as a random holonomy. For the construction of this *kinematical measure*, see [5, 4] and also [7], where the authors present general motivations for a theory of integration over connections.

This generalizes the previous definition of $\mathrm{hol}(\omega, l)$, since the *joint conjugacy class* of $(g_1, \ldots, g_n)$, as we shall call its orbit under the adjoint action of $G$, determines the individual conjugacy classes of $g_1, \ldots, g_n$. In [36], A. Sengupta has proved the following result[4].

PROPOSITION 1. *Let $\omega$ and $\omega'$ be two connections. If for all $n$-tuple $l_1, \ldots, l_n$ of piecewise smooth loops on $M$ one has the equality*

$$\mathrm{hol}(\omega, l_1, \ldots, l_n) = \mathrm{hol}(\omega', l_1, \ldots, l_n) \tag{14}$$

*in $G^n/\mathrm{Ad}$, then $\omega$ and $\omega'$ are equivalent modulo gauge transformations.*

It is not clear whether it is always the case that the values of the Wilson loops determine quantities like $\mathrm{hol}(\omega, l_1, \ldots, l_n)$. However, another result by A. Sengupta, proved in [37], shows that this is true when $G$ is one of the groups $U(n)$, $SU(n)$, $SO(2n+1)$ or a product of some of them, or one of the groups $Spin(2n+1)$, $Pin(n)$, $SO(4)$, $Spin(4)$. This covers for example the case of $U(1) \times SU(2) \times SU(3)$ which is the group involved in the standard model.

An important consequence of Proposition 1 is the possibility of imbedding the quotient $\mathcal{A}/\mathcal{J}$ in a quotient of a space of functions. Indeed, let $LM$ be a class of loops on $M$, say piecewise smooth. The space $\mathcal{F}(M, G)$ of all $G$-valued functions on $M$ acts on the space $\mathcal{F}(LM, G)$ of $G$-valued functions on $LM$ by

$$(\phi \cdot f)(l) = \phi(l_0)^{-1} f(l) \phi(l_0). \tag{15}$$

Now one can reformulate Proposition 1 as follows.

PROPOSITION 2. *The mapping induced by the holonomy*

$$\mathrm{hol} : \mathcal{A}/\mathcal{J} \longrightarrow \mathcal{F}(LM, G)/\mathcal{F}(M, G) \tag{16}$$

*is an injection.*

The point of view of a *random holonomy* rather than a *random connection* which is adopted in most of the works on the Yang-Mills measure, including the present one, is legitimated at an informal level by this result[5]. Indeed the space of generalized holonomies, as one might call $\mathcal{F}(LM, G)/\mathcal{F}(M, G)$, is an enlargement of the space of smooth connections modulo gauge transformations. From a probabilistic perspective, this puts the problem in a familiar context, that of probability measures on function spaces.

We are now going to present an infinite-dimensional approach to the construction of the measure which relies on a similarity between (3) and the definition of a Gaussian measure.

### Infinite-dimensional approach

From now on, let us concentrate on the case where $M$ is two-dimensional and orientable. An important feature of the two-dimensional Yang-Mills action is its invariance under the diffeomorphisms which preserve the Riemannian volume. Actually, we can forget the Riemannian structure, and endow $M$ with a volume form

---

[4] A. Sengupta has pointed out that, in the proof, the second usage of the word "any" should be replaced by "the same".

[5] It should be noted that D. Fine [22, 23] has used a different approach. He has taken advantage from a careful investigation of the geometric structure of the quotient $\mathcal{A}/\mathcal{J}$ to construct the measure directly on it.

$\sigma$, or, if we do not need to choose an orientation, with the corresponding density. Then the action can be rewritten as

$$S(\omega) = \int_M \|F\|^2 \, d\sigma. \tag{17}$$

This has at least two consequences. The first one is that the theory is invariant under a large group of diffeomorphisms, making two-dimensional Yang-Mills theory look similar to a topological theory. This will be illustrated at the end of Chapter 5 through the study of partition functions.

The second consequence concerns the case of an Abelian structure group, that is, $G = U(1)^n$. In such a group, the quadratic relation (6) becomes

$$F = dA \tag{18}$$

which is *linear*. This suggests a change of variable in (3), leading to the Gaussian-like expression

$$d\mu(F) = \frac{1}{Z} e^{-\frac{1}{2}\|F\|_{L^2}^2} \, dF. \tag{19}$$

This heuristic argument is very appealing because it suggests the following interpretation of (3): *under the Yang-Mills measure, the curvature of a random connection has a Gaussian distribution*. This interpretation proves very helpful in building one's intuition of the measure when $G$ is Abelian.

In the simple case of a loop $l$ bounding a domain $D$, Stoke's formula allows us to compute the holonomy along $l$ given the curvature inside $D$, by

$$\text{hol}(\omega, l) = \exp \int_l A = \exp \int_D F. \tag{20}$$

It seems reasonable to replace $F$ in this formula by an $i\mathbb{R}^n$-valued Gaussian random $L^2$ function on $(M, \sigma)$, that is, a *white noise* $W$ on $M$ with intensity $\sigma$. This leads to the definition of a $U(1)^n$-valued random variable

$$H_l = \exp \int_D W \, d\sigma, \tag{21}$$

disregarding the fact that $W$ is not a function but a distribution.

What gives its full strength to this discussion is that it can be extended to the case of an arbitrary group. Indeed, it is locally possible, given a connection $\omega$, to privilege a direction on $M$ and to choose a local section $s$ of $P$ such that $A = s^*\omega$ vanishes in this direction. Then, since $M$ is two-dimensional, this implies that the quadratic part $[A, A]$ of (6) vanishes identically. Through this particular section, or in this particular *axial gauge*, $F$ depends linearly on $A$.

This observation is at the starting point of the work of A. Sengupta [36, 38] on general surfaces following that of B. Driver [21, 20] on the plane. His strategy allows him to define the random holonomy along a class of loops wide enough to characterize a connection in the sense of Proposition 1, and to compute the finite-dimensional marginals of this process, for any particular type of principal bundle.

However, the class of loops considered in [38] is strongly dependent on the choice of the privileged direction and it is not invariant under diffeomorphisms.

Roughly speaking, the problem one encounters when computing the holonomy along the boundary of a disk by replacing the curvature by a Lie-algebra valued white noise is to determine in which order one should multiply the small elements of $G$ obtained by exponentiating the integrals of the white noise over small elementary domains inside the disk. Local coordinates allow one to choose such an order in a consistent way but only along those loops that A. Sengupta named *admissible*. As such, this problem has no intrinsic solution. However, L. Gross proposed to replace the curvature as the central object of interest by what he called *lasso forms*. He showed how Stokes' formula and the Poincaré lemma can be generalized in this context by using these new objects (see [27] for a survey and [26] for a detailed account). Although B. Driver has studied lasso forms in relation with the measure in [20], there remains a lot to understand in that direction.

## Finite-dimensional approach

Our approach of the construction goes in the reverse direction. We start from the finite-dimensional marginals as they are described in the physical literature, build them into a consistent discrete theory and then take the continuous limit of this discrete theory. In this way, we are able to define the random holonomy along the class of piecewise embedded loops, which is stable under area-preserving diffeomorphisms.

Let us present in more detail our approach by going through the content of each chapter[6].

Chapter 1 is a complete and self-contained exposition of the discrete theory. Discretizing space-time means essentially replacing $M$ by a graph and considering only the loops or paths that can be traced in this graph. To such a discretization one can associate a probability measure on a finite-dimensional manifold, namely $G^r$ where $r$ is the number of edges of the graph. The heat kernel on $G$ plays a central role in the discrete theory and it was first introduced – if implicitly – by A. Migdal in 1975 [32]. The discrete theory was then also discussed by S. Albeverio et al. in [2, 3], by B. Driver in [20] and by E. Witten in [44].

Chapter 2 is the heart of this work. It is devoted to taking the continuous limit of the discrete theory. This is difficult for a simple reason: if one superposes two finite graphs on $M$, the result is not necessarily a finite graph, because two smooth curves can intersect badly[7]. Put in other words, the discrete theory does not provide all finite-dimensional marginals of the random holonomy process. However, we show that one can approximate in a proper way any finite family of loops by families for which the discrete theory is effective. This leads us to the construction of a stochastic process of random holonomy indexed by all piecewise embedded paths, with the possibility of putting constraints on the holonomy along a finite number of non-intersecting simple loops [8]. We prove the existence of a pathwise multiplicative version of this process, in a way which is related to the use of projective techniques by A. Ashtekar and J. Lewandowski in their work on the kinematical measure [5, 6]. As far as I know, this is one of the first instances of a pathwise property of the random holonomy process.

---

[6]The reader will notice that there is no mention of the small-volume limit of the theory in the present work. We address him to [39] and the references therein.

[7]J. Baez and S. Sawin [9, 10] have analyzed the possible pathology of the intersection of piecewise immersive curves and introduced *webs*, a generalization of the ynotion of graph which allows to take such pathology into account. However, a discrete Yang-Mills theory on webs is not yet available

[8]The general construction presented in this memoir does however not take the particular topology of the bundle into account, and we expect to fill this gap in the future. See however the description of Chapter 3.

Chapter 3 is devoted to the case of an Abelian group, the case where the measure is essentially Gaussian. We show that one can go from a white noise to the Yang-Mills measure and vice versa. The topology of the fiber bundle appears here explicitely in the computations and we are in this special case able to construct the random holonomy corresponding to a particular topological type of bundle.

Chapter 4 contains the proof of a result which had been announced earlier [**31**]. It points out a major difference in the local structure of the measure between the cases of an Abelian and a semi-simple group $G$. In particular, it shows that the procedure used to extract a white noise from the measure in Chapter 3 would, when $G$ is semi-simple, produce a deterministic object. This result is closely related to the problem discussed above of determining to what extent the Yang-Mills measure can be seen as a functional of a white noise. It reveals in fact a property of non-locality of the random field which is not yet completely understood.

In Chapter 5 we investigate the behaviour of the measure under the surgery of surfaces. The Markov property of the Yang-Mills field was already known, it had been studied for example in [**2**] and proved in [**12**]. We have not really concentrated our effort on getting the optimal version of the Markov property, but rather on understanding a phenomenon which is related to the non-locality mentionned above. We show that the two sub-$\sigma$-algebras generated by the Yang-Mills field on two halves of a surface do not generate the full ambient $\sigma$-algebra and we identify the defect of information. Finally, we summarize the properties of the partition functions proved at various stages of the study and organize them as to emphasize the close relationship between the Yang-Mills theory and a topological quantum field theory.

---

Yves Le Jan supervised the PhD thesis of which this monography is a revised version and I would like him to find here an expression of my deepest gratitude. Part of the revision work has been possible thanks to the kind hospitality of the Statistical Laboratory in Cambridge (UK) where I have benefitted from fruitful discussions with James Norris.

CHAPTER 1

# Discrete Yang-Mills measure

In this first chapter, we shall construct and study the discrete Yang-Mills measure. This requires quite a long introductory part to set up the notation, explain how graphs are used - for a while - instead of surfaces and how the concepts of bundle, connection, holonomy and gauge transformation can be adapted to that framework. This will be achieved by the end of Section 1.3. The discrete Yang-Mills measure and its conditional versions are then defined in Sections 1.4 and 1.5 respectively. The two next sections are devoted to the proof of two fundamental properties of the measure, namely its invariance under subdivision and its invariance under the action of area-preserving diffeomorphisms. In Section 1.8, we translate a classical small-time estimate for the heat kernel on a compact Lie group into a fundamental estimate for the holonomy along small loops. Finally, Section 1.9 presents the first part of the study of the special case $G = U(1)$ which will be completed in Chapter 3.

It seems that the discrete Yang-Mills measure had not yet been given as autonomous and complete a treatment as it is given here. Nevertheless, gauge theories on lattices are extensively used and studied in the physical litterature, and, since the discrete setting is the natural framework for computations, it also appears in a number of mathematical works, for example those of B. Driver [20] and A. Sengupta [36, 38]. Another important source of inspiration has been the paper [44] where E. Witten outlined a discrete theory in order to compute the volume of moduli spaces of flat connections.

## 1.1. Notation

Throughout this work, $M$ will denote a surface, that is, a real differentiable two-dimensional manifold, compact, connected, orientable, with or without boundary. This means that $M$ is a sphere, a torus, or a surface of higher genus, from which a finite number of open disks have possibly been removed. This surface is endowed with a Lebesguian measure $\sigma$, a measure which has a positive smooth density with respect to the Lebesgue measure in any coordinate chart. We could just as well take a non-vanishing smooth differential 2-form on $M$ but this would induce a choice of orientation, that we do not need in general.

The boundary of $M$, if not empty, is the disjoint union of some circles $N_1, \ldots, N_p$. As usual in such a situation, we want to call *smooth* the objects defined on $M$ which are the restriction of smooth objects defined on an open neighbourhood of $M$. The notion of closure will make this statement more precise and will also prove very useful in the next chapter when we will be constructing the continuous Yang-Mills measure.

DEFINITION 1.1. A *closure* of $M$ is a triple $(i, M, M_1)$, where $M_1$ is a closed surface, i.e. a surface without boundary and $i : M \longrightarrow M_1$ is an embedding. If the complement of $i(M)$ in $M_1$ is diffeomorphic to a disjoint union of disks, the closure is said to be *minimal*.

Given two closures $(i_1, M, M_1)$ and $(i_2, M, M_2)$ of $M$, $i_1(M)$ and $i_2(M)$ have diffeomorphic neighbourhoods in $M_1$ and $M_2$ and it makes sense to say that a mapping (resp. a bundle, a section,...) is smooth on $M$ if it is the restriction of a smooth mapping (resp. bundle, section,...) defined on an open neighbourhood of $M$ in one of its closures.

The other basic object we need is a compact connected Lie group $G$, that will be chosen to be Abelian or semi-simple in most examples. In fact there would be little harm in considering that $G$ is one of the groups $U(1)$, $SU(2)$ and $SO(3)$, or a product of some of these groups.

Given $M$ and $G$, we consider a principal $G$-bundle $P$ over $M$. For a definition of a principal bundle and various related objects that we are going to use, the reader could refer for example to [**30**] or [**13**][1]. The following topological considerations are classical.

If $M$ has a boundary, then $M$ retracts on a bunch of circles so that $P$ is trivial, that is, it is diffeomorphic over $M$ to the product $M \times G$. If $M$ is closed, then $P$ does not necessarily have this nice property and the possible topological structures for $P$ are classified by $\pi_1(G)$. A pleasant way to see this is to cut $M$ along the boundary of a small disk, producing two disjoint pieces. The restrictions of a bundle $P$ over $M$ to both pieces are trivial and the topology of $P$ is completely determined by the transition function along the boundary of the disk, which tells us how these restrictions are sewed together. This transition function is a map $S^1 \longrightarrow G$ and homotopic maps give rise to equivalent bundles. If $G = SU(2)$ for example, then $P$ is necessarily trivial, because $\pi_1(SU(2)) = 0$. Note also that when $G$ is semi-simple, $\pi_1(G)$ is a finite group.

A connection[2] $\omega$ on $P$ is a choice at each point $p$ of $P$ of a subspace of the tangent space $T_p P$ supplementary to the vertical subspace, where vertical means tangent to the fibre through $p$. Moreover, this distribution of subspaces, called the *horizontal distribution*, must be invariant under the action of $G$.

Let $c : [0, 1] \longrightarrow M$ be a piecewise $C^1$ path on $M$. A connection $\omega$ allows one to lift $c$ to a horizontal path in $P$ starting at any prescribed point in $P_{c(0)}$, the fibre over $c(0)$. The mapping which sends a point $p$ of $P_{c(0)}$ to the endpoint of the horizontal lift of $c$ starting at $p$ is called the parallel transport or holonomy of $\omega$ along $c$. It is a $G$-equivariant map $\mathrm{hol}(\omega, c) : P_{c(0)} \longrightarrow P_{c(1)}$ in that $\mathrm{hol}(\omega, c)(p.g) = \mathrm{hol}(\omega, c)(p).g$. If $c_1$ and $c_2$ are two paths such that $c_1(1) = c_2(0)$, then the concatenated path $c_1 c_2$

---

[1]Footnotes in the first sections are intended to provide the reader at least with a rough understanding of the geometrical picture, in case he needs it. For instance, the canonical example of a principal $G$-bundle over $M$ is the direct product $M \times G$, on which $G$ acts by right multiplication on the second factor.

[2]A connection associates to each piecewise $C^1$ path on $M$ its *parallel transport* or *holonomy*, which is, in first approximation, an element of $G$. Formulae (1.1) and (1.2) for the composition of paths and for the action of a gauge transformation can be taken as axioms.

exists and we have the following relation :
$$\mathrm{hol}(\omega, c_1 c_2) = \mathrm{hol}(\omega, c_2) \circ \mathrm{hol}(\omega, c_1). \tag{1.1}$$

A gauge transformation[3] is a diffeomorphism $\phi$ of $P$ over the identity of $M$ which commutes to the right action of $G$. Let $\omega$ be a connection on $P$. A gauge transformation $\phi$ allows one to define a new connection $\phi^*\omega$ whose holonomy is related to that of $\omega$ by the relation:
$$\mathrm{hol}(\phi^*\omega, c) = (\phi_{|P_{c(1)}})^{-1} \circ \mathrm{hol}(\omega, c) \circ \phi_{|P_{c(0)}}, \tag{1.2}$$
for each piecewise $C^1$ path $c$. Note that these holonomies are *conjugated* if $c(0) = c(1)$, i.e. if $c$ is a loop[4].

## 1.2. Discretization of a surface

The role of graphs is to allow us to replace the infinite-dimensional space of connections modulo gauge transformations by a finite-dimensional manifold.

### 1.2.1. Pregraphs.
To stay consistent with our previous convention, we say that an application $c : [0,1] \longrightarrow M$ is smooth (resp. an embedding) if it is the restriction of a smooth application (resp. an embedding) defined on an open interval containing $[0,1]$.

DEFINITION 1.2. A *parametrized path* on $M$ is a mapping $c : [0,1] \longrightarrow M$ which is either constant or the concatenation of a finite number of smooth embeddings.

Each time we will be talking about concatenation of two paths, we will implicitly assume that the endpoint of the first one is the starting point of the second. Moreover, we will always parametrize non-constant paths in such a way that they are constant on no interval.

We say that two parametrized paths are equivalent if they differ by an increasing piecewise smooth reparametrization. This equivalence relation preserves orientation, image, endpoints, injectivity and injectivity on $(0,1)$ of the parametrized paths. Equivalence classes of parametrized paths will be called simply *paths*. The set of paths on $M$ will be denoted by $PM$. Observe that $PM$ is stable under the diffeomorphisms of $M$, in that that the image of any path by a diffeomorphism is still a path. The reason why this matters is that the Yang-Mills measure ought to be invariant under area-preserving diffeomorphisms.

A path whose endpoints are equal is a *loop* and a loop which is injective on $[0,1)$ is said to be *simple*. Given a path $c$, we denote by $c^{-1}$ the path obtained by reversing the orientation of $c$. An *edge* is an injective path $e$ such that $e([0,1]) \cap \partial M$ is either empty or a finite union of segments.

DEFINITION 1.3. A *pregraph* on $M$ is a set $\Gamma = \{e_1, \ldots, e_r\}$ of edges which meet each other only at their endpoints, i.e. such that for all distinct $i$ and $j$ between 1 and $r$, one has
$$e_i([0,1]) \cap e_j([0,1]) = e_i(\{0,1\}) \cap e_j(\{0,1\}).$$

---

[3] A gauge transformation on $M \times G$ is a map $\phi : M \longrightarrow G$ which acts on $M \times G$ by $\phi \cdot (m,g) = (m, \phi(m)g)$. Note that it acts on the left, whereas $G$ acts on the right: these actions commute.

[4] It should be kept in mind that a connection associates in an intrinsic way a *conjugacy class* of $G$ to each loop on $M$. See the introduction for a more detailed discussion.

We call *support* of a pregraph $\Gamma$ the union of the images of its edges. A pregraph $\Gamma$ is said to be connected if its support $Supp(\Gamma)$ is connected.

We call *faces* of a pregraph $\Gamma$ the connected components of $M \setminus Supp(\Gamma)$ and denote by $\mathcal{F}(\Gamma)$ the set of these faces.

We want to consider special pregraphs, those which encode the full topology of $M$. A convenient measure of the "amount of topology" is provided by the homology groups and the following proposition gives us a simple sufficient condition in this respect.

PROPOSITION 1.4. *Let $\Gamma$ be a connected pregraph on $M$. Suppose that each face of $\Gamma$ is diffeomorphic to a disk. Then the map $H_1(Supp(\Gamma); \mathbb{Z}) \longrightarrow H_1(M; \mathbb{Z})$ induced by the inclusion is surjective.*

PROOF. Let $l : [0,1] \longrightarrow M$ be a loop. There exists on each face of $\Gamma$ a point which is not in the image of $l$. In each face, fix such a point and remove it from $M$. The remaining open set $U$ retracts on the support of $\Gamma$ because each punctured closed face retracts on its boundary. (To see this, observe that the closure of each face of $\Gamma$ is the image of a closed disk by a continuous mapping which is a homeomorphism on the interior of the disk.) This retraction induces a homotopy from $l$ onto a loop in $Supp(\Gamma)$. So each loop of $M$ is homotopic, thus homologous to a loop of $Supp(\Gamma)$. This proves the result. □

Connectedness seems also to be a sensible condition to impose to a graph.

LEMMA 1.5. *A pregraph whose faces are diffeomorphic to disks is necessarily connected.*

PROOF. Let $\Gamma$ be a pregraph and assume that each face of $\Gamma$ is diffeomorphic to a disk. Suppose that $\Gamma = \Gamma' \cup \Gamma''$, where $\Gamma'$ and $\Gamma''$ have non-empty disjoint supports. Each face of $\Gamma$ is a disk, so it has a connected boundary, which is included either in the support of $\Gamma'$ or in that of $\Gamma''$. The unions of the closures of the faces whose boundaries lie in $Supp(\Gamma')$ (resp. $Supp(\Gamma'')$) partition $M$ into two non-empty closed sets, contradicting its connectedness. □

Finally, we would like to make sure that the boundary of $M$, if not empty, is taken into account. This requires the definition of a path in a graph.

By a *path* in a pregraph $\Gamma$ we mean a concatenation of edges of $\Gamma$, with natural or reversed orientation, and we will denote by $\Gamma^*$ the set of these paths. We are now in position to define *graphs*.

### 1.2.2. Graphs.

DEFINITION 1.6. *A graph on $M$ is a connected pregraph $\Gamma$ whose faces are diffeomorphic to disks and such that for each component $N_i$ of $\partial M$, there exists an element of $\Gamma^*$ whose image is equal to $N_i$.*

Recall that connectedness is a consequence of the faces being diffeomorphic to disks.

DEFINITION 1.7. *Let $\Gamma_1$ and $\Gamma_2$ be two pregraphs. We say that $\Gamma_2$ is finer than $\Gamma_1$ and write $\Gamma_1 \leq \Gamma_2$ if each edge of $\Gamma_1$ is a path of $\Gamma_2^*$.*

PROPOSITION 1.8. *Any pregraph can be refined into a graph with the same number of faces.*

Before proving Proposition 1.8, let us review some classical facts about the topology of $M$. If $M$ has no boundary and is not a sphere, its universal covering is diffeomorphic to a plane. In this plane, it is always possible to choose a polygonal fundamental domain for the covering map, namely a $4g$-gonal domain if $g$ is the genus of $M$. This means that it is possible to see $M$ topologically as a polygon in which one has identified certain edges. If $M$ is a shpere, it can be seen as a disk whose upper and lower half-boundary have been identified.

If $M$ has a boundary, then we can see it as a closed surface from which some disks have been removed and it is possible to choose a polygonal representation of the closed surface such that the removed disks lie inside the interior of the polygon. Figure 1 illustrates this point.

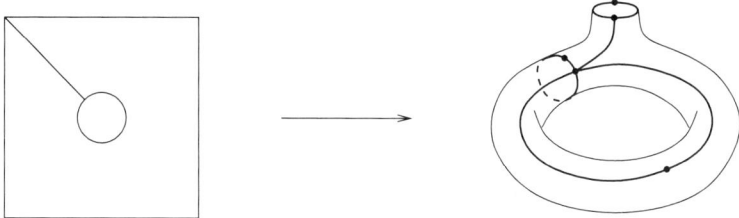

FIGURE 1. A graph with one face on a torus with one hole.

PROOF OF PROPOSITION 1.8. Let $\Gamma_0$ be a pregraph. We do not assume that $\Gamma_0$ is non-empty. By definition of an edge, the set $Supp(\Gamma_0) \cap \partial M$ is a finite union of segments. By cutting some edges of $\Gamma_0$ into several pieces if necessary, we may assume that these segments are exactly the images of some edges of a pregraph $\Gamma_1$ finer than $\Gamma_0$ and with the same support. Then, it is possible to add new edges to $\Gamma_1$ in such a way that $Supp(\Gamma_1) \cap \partial M = \partial M$. Observe that, if $Supp(\Gamma_1)$ did not meet some component of $\partial M$ initially, it is necessary to add at least two edges on this component, for an edge is injective by definition.

At this stage, each face of $\Gamma_1$ is homeomorphic to the interior of a compact surface with boundary. There remains only to see that on every surface there exists a graph with exactly one face. If the surface is closed, then the boundary of a fundamental domain in the universal covering projects on a system of loops on $M$. Cutting each of these loops in two edges produces a graph with only one face. If the surface has a boundary, one needs just add three edges for each boundary component, two covering it and one joining it to the boundary of the fundamental domain. □

We will use several times the fact that there exists a graph with exactly one face on any surface $M$.

## 1.3. Discrete holonomy and gauge transformations

We want to adapt the notions of holonomy and gauge transformation to the discrete setting. Let us fix a graph $\Gamma = \{e_1, \ldots, e_r\}$ on $M$ and restrict our attention to the parallel transport along the paths of $\Gamma^*$, those obtained by concatenation of edges of $\Gamma$.

From a topological point of view, restricting ourselves to $\Gamma$ makes us lose the particular structure of $P^5$. More precisely, denoting by $i : Supp(\Gamma) \longrightarrow M$ the inclusion map, we are saying that the restricted bundle $i^*P$ over $Supp(\Gamma)$ is always trivial, because $G$ is connected. We will discuss the significance of this loss later but for the moment it allows us, by using a section of $i^*P$, to represent the holonomies along the edges of $\Gamma$ by elements of $G$. For example, after choosing an identification of $i^*P$ with $Supp(\Gamma) \times G$, we can define $h_c(\omega)$ in $G$ representing the holonomy along a path $c$ by the relation $hol(\omega, c)(c(0), 1) = (c(1), h_c(\omega))$.

Looking at the relation (1.1), we see that $h_c$ is equal to $(h_{e_n})^{\varepsilon_n} \ldots (h_{e_1})^{\varepsilon_1}$ if $c = e_1^{\varepsilon_1} \ldots e_n^{\varepsilon_n}$, i.e. if $c$ is the concatenation in this order of $e_1^{\varepsilon_1}, \ldots, e_n^{\varepsilon_n}$, with $\varepsilon_i = \pm 1$. Once we are convinced that the decomposition of a path of $\Gamma^*$ into a concatenation-product of edges is unique, we can set the following definition.

DEFINITION 1.9. By a *discrete connection* on $\Gamma$ we mean a mapping from $\Gamma$ into $G$. The set of discrete connections will be denoted by $G^\Gamma$ and given an element $g = (g_e)_{e \in \Gamma}$ of $G^\Gamma$, we define the *discrete holonomy* of $g$ along a path $c$ of $\Gamma^*$ by

$$h_c(g) = (g_{e_n})^{\varepsilon_n} \ldots (g_{e_1})^{\varepsilon_1}, \tag{1.3}$$

where $c = e_1^{\varepsilon_1} \ldots e_n^{\varepsilon_n}$ is the unique decomposition of $c$ as a concatenation of edges. Thus, for all $c \in \Gamma^*$, $h_c$ is a $G$-valued function on $G^\Gamma$.

The following property of the discrete holonomy is obvious but it is so important that it is worth stating separately. It will be refered to as the *multiplicativity* of the holonomy.

PROPOSITION 1.10. *Let $c_1$ and $c_2$ be two paths of $\Gamma^*$ such that the concatenation $c_1 c_2$ exists. Then we have the equality of functions*

$$h_{c_1 c_2} = h_{c_2} h_{c_1}.$$

Ideally, a connection on $i^*P$ should allow us to compute more than only the parallel transport along an edge or a product of edges. We should be able to take into account any path in $Supp(\Gamma)$. But we are interested in connections *modulo gauge transformations* and the following lemma tells us that our space of discrete connections is big enough. We introduce the notation $\mathcal{V}(\Gamma)$ for the set of vertices of $\Gamma$.

LEMMA 1.11. *Let $\omega_1$ and $\omega_2$ be two connections on $i^*P$. Suppose that $\omega_1$ and $\omega_2$ have the same holonomy along each edge of $\Gamma$. Then there exists a gauge transformation $\phi$ of $i^*P$ which leaves invariant the fibres over the points of $\mathcal{V}(\Gamma)$ and such that $\phi^*\omega_1 = \omega_2$.*

PROOF. Choose a parametrization of an edge $e \in \Gamma$. Define a gauge transformation over $e$ by

$$\phi(p) = hol(\omega_1, e_{|[0,t]}) \circ hol(\omega_2, (e_{|[0,t]})^{-1})(p) \quad \forall p \in P_{e(t)}, \; t \in [0,1].$$

Then

$$hol(\omega_2, e_{|[0,t]}) = \phi^{-1} \circ hol(\omega_1, e_{1|[0,t]}) \circ \phi = hol(\phi_e^* \omega_1, e_{1|[0,t]}),$$

using (1.2) and the fact that $\phi_{|P_{e(0)}} = Id_{P_{e(0)}}$. Since $\omega_1$ and $\omega_2$ have the same holonomy along $e$, $\phi$ leaves the fibre over the endpoint of $e$ invariant. Doing this

---

[5] The geometrical content of this section can be ignored without much inconvenience. What one really needs to read before proceeding with the next section is Definition 1.9, Proposition 1.10, Definition 1.12 and the comments following it.

construction over each edge of $\Gamma$ provides us with a well-defined gauge transformation $\phi$ of $i^*P$ such that $\omega_2$ and $\phi^*\omega_1$ have the same holonomy, hence the same horizontal paths and are in fact equal. $\square$

The converse to Lemma 1.11 follows immediately from (1.2), so that $G^\Gamma$ is actually the space of connections on $i^*P$ modulo those gauge transformations that leave the fibres over $\mathcal{V}(\Gamma)$ invariant. It remains to take into account the action of a general gauge transformation on these fibres.

DEFINITION 1.12. By a *discrete gauge transformation* on $\Gamma$ we mean a mapping from $\mathcal{V}(\Gamma)$ into $G$. The set of discrete gauge transformations will be denoted by $G^{\mathcal{V}(\Gamma)}$ and any element $\phi = (\phi_v)_{v \in \mathcal{V}(\Gamma)}$ of $G^{\mathcal{V}(\Gamma)}$ acts on the set of discrete connections by

$$\phi \cdot g = ((\phi \cdot g)_e)_{e \in \Gamma} = ((\phi_{e(1)})^{-1} g_e \phi_{e(0)})_{e \in \Gamma} \,, \quad g \in G^\Gamma. \tag{1.4}$$

A discrete gauge transformations also acts by duality on any function defined on $G^\Gamma$. In particular, a gauge transformation $\phi$ makes the discrete holonomy $h_c : G^\Gamma \longrightarrow G$ into $h_c \circ \phi$, with

$$h_c \circ \phi = (\phi_{c(1)})^{-1} h_c \phi_{c(0)} \,, \quad c \in \Gamma^*. \tag{1.5}$$

The most important consequence of this formula is that *the conjugacy class of the holonomy along a loop is invariant under gauge transformations*. This is the key to the construction of the discrete Yang-Mills measure, as we will see right now.

### 1.4. Discrete Yang-Mills measure

Given a graph $\Gamma = \{e_1, \ldots, e_r\}$ on $M$, we are now keen to replace the infinite-dimensional space $\mathcal{A}/\mathcal{J}$ of connections modulo gauge transformations on $P$ by the quotient of the finite-dimensional manifold $G^\Gamma$ by the action of $G^{\mathcal{V}(\Gamma)}$ and to look for the discrete Yang-Mills measure as a probability measure on $G^\Gamma$ invariant under discrete gauge transformations.

The basic example of such an invariant measure is the product of Haar measures and we shall construct the discrete Yang-Mills measure $P$ on $G^\Gamma$ under the form:

$$dP = \frac{dP}{dg} dg,$$

where $dg = dg_1 \otimes \ldots \otimes dg_r$ is the product of the unit-mass Haar measures on each factor. The density $\frac{dP}{dg}$ will be chosen as a product of central functions of holonomies along loops, a feature that will make $P$ gauge-invariant. Recall that a function $p$ is said to be central on $G$ if $p(xy) = p(yx)$, or equivalently if $p(yxy^{-1}) = p(x)$ for all $x, y$ in $G$.

The boundary of each face $F$ of $\Gamma$ is the image of a path defined up to the choice of an origin and an orientation. The function $h_{\partial F} : G^\Gamma \longrightarrow G$ is thus defined only up to conjugation and inversion, but this is enough for $p \circ h_{\partial F}$ to be well-defined whenever $p$ is a central function invariant by inversion.

Let us endow $G$ with its bi-invariant Riemannian metric with total volume equal to 1 and denote by $(p_t)_{t>0}$ the fundamental solution of the heat equation on $G$. It satisfies

$$(\partial_t - \frac{1}{2}\Delta)p_t = 0 \quad \text{on} \quad \mathbb{R}_+^* \times G, \tag{1.6}$$

and for any function $f$ continuous on $G$,

$$\int_G f(g)p_t(g)\,dg \xrightarrow[t\to 0]{} f(1), \tag{1.7}$$

where 1 denotes the unit element in $G$. For any positive $t$, $p_t$ is a positive central function, invariant by inversion, such that $\int_G p_t(g)\,dg = 1$.

According to our previous discussion, the function $p_{\sigma(F)}(h_{\partial F}) : G^\Gamma \longrightarrow \mathbb{R}_+^*$, where $\sigma$ denotes the surface measure on $M$, is well-defined for each face $F$ of $\Gamma$, so that it makes sense to define

$$D = \prod_{F \in \mathcal{F}(\Gamma)} p_{\sigma(F)} \circ h_{\partial F} : G^\Gamma \longrightarrow \mathbb{R}_+^*, \tag{1.8}$$

$$Z = \int_{G^\Gamma} D\,dg, \tag{1.9}$$

and finally to define $P$ on $(G^\Gamma, \mathcal{B}(G^\Gamma))$ by

$$dP = \frac{1}{Z} D\,dg. \tag{1.10}$$

This measure is called the *discrete Yang-Mills measure* on $\Gamma$. That it is gauge invariant is obvious but it will be stated and proved properly in Lemma 1.18.

The choice of the heat kernel may seem to be quite arbitrary, and it is, to a certain extent. We shall discuss this point at the end of Section 1.6.

Given a finite collection paths $c_1, \ldots, c_n$ in $\Gamma^*$, we can now define the *law of the discrete holonomy* along $c_1, \ldots, c_n$ as the joint distribution of the $n$-tuple $(h_{c_1}, \ldots, h_{c_n})$ under $P$. This law is a probability measure on $G^n$.

Before studying the measure in greater detail, we shall extend slightly its definition in order to deal with boundary conditions.

### 1.5. Conditional Yang-Mills measure

When $M$ has a boundary, one expects to be able to impose boundary conditions on the holonomy. It will also prove useful to be able to impose conditions on the holonomy along other loops than the boundary components, even when $M$ has no boundary.

#### 1.5.1. Conditional Haar measure.
Let us start with an easy disintegration result for the reference measure.

PROPOSITION 1.13. *Let $n$ be a positive integer. Let $x$ be an element of $G$. There exists on $G^n$ a unique measure $\nu_x^n$ such that $g_n \ldots g_1 = x$ $\nu_x^n$-a.s. and such that for any function $f$ continuous on $G^n$ and any $i$ between 1 and $n$,*

$$\nu_x^n(f) = \int_{G^{n-1}} f(g_1, \ldots, g_{i-1}, (g_n \ldots g_{i+1})^{-1} x (g_{i-1} \ldots g_1)^{-1}, g_{i+1}, \ldots, g_n)$$
$$dg_1 \ldots \widehat{dg_i} \ldots dg_n.$$

*Moreover, one has*

$$\nu_x^n(f) = \lim_{t\to 0} \int_{G^n} f(g_1, \ldots, g_n)\, p_t(g_n \ldots g_1 x^{-1})\,dg_1 \ldots dg_n.$$

*Finally, $\int_G \nu_x^n\,dx = dg$ holds as an equality of measures on $G^n$.*

PROOF. Pick $i$ between 1 and $n$, and $t > 0$. By centrality of $p_t$ and then right invariance of $dg_i$, one has

$$\int_{G^n} f(g_1,\ldots,g_n) p_t(g_n \ldots g_1 x^{-1}) \, dg =$$

$$= \int_{G^n} f(g_1,\ldots,g_n) \, p_t(g_i(g_{i-1} \ldots g_1) x^{-1} (g_n \ldots g_{i+1})) \, dg$$

$$= \int_{G^n} f(g_1,\ldots,g_{i-1}, g_i(g_n \ldots g_{i+1})^{-1} x(g_{i-1} \ldots g_1)^{-1}, g_{i+1},\ldots,g_n) \, p_t(g_i) \, dg_i$$

$$dg_1 \ldots \widehat{dg_i} \ldots dg_n$$

$$\xrightarrow[t \to 0]{} \int_{G^{n-1}} f(g_1,\ldots,g_{i-1},(g_n \ldots g_{i+1})^{-1} x(g_{i-1} \ldots g_1)^{-1}, g_{i+1},\ldots,g_n)$$

$$dg_1 \ldots \widehat{dg_i} \ldots dg_n.$$

Thus the limit exists and the last expression does not depend on $i$. It defines a probability measure on $G^n$, whose uniqueness is obvious.

If $f$ vanishes on the hypersurface $\{g_n \ldots g_1 = x\}$, then $\nu_x^n(f) = 0$, so we do have $g_n \ldots g_1 = x$ $\nu_x^n$-almost surely.

Finally, since $\int_G p_t(g) \, dg = 1$, we have for any $t > 0$

$$\int_G \int_{G^n} f(g_1,\ldots,g_n) \, p_t(g_n \ldots g_1 x^{-1}) \, dx dg = \int_{G^n} f(g_1,\ldots,g_n) \, dg,$$

which implies the last statement when $t$ tends to zero. □

### 1.5.2. Conditional Yang-Mills measure.
Suppose we are given $q$ loops $L_1,\ldots,L_q$ and that we want to impose conditions on the holonomy along each of these loops. To keep things computable, we will assume that they are simple loops and that they do not meet each other. We make the further assumption that each of them is either fully contained in a boundary component or fully contained in the interior of $M$. Note that in the first case, the loop goes exactly once around the boundary component.

We are now looking for a disintegration of $P$ with respect to $(h_{L_1},\ldots,h_{L_q})$.

Let $\Gamma'$ denote the set of edges of $\Gamma$ which do not appear in the decomposition of any $L_i$. We will denote by $dg'$ the product of unit-mass Haar measures on $G^{\Gamma'}$. The definition of the conditional measure is very natural but the fact that the conditional Haar measure is not invariant by permutation of the factors on $G^n$ forces one to adopt a slightly elliptic notation.

Pick a point $(x_1,\ldots,x_q)$ in $G^q$. Suppose that the edges are labeled in such a way that $L_1 = e_1^{\varepsilon_1} \ldots e_{n_1}^{\varepsilon_{n_1}}, \ldots, L_q = e_{n_{q-1}+1}^{\varepsilon_{n_{q-1}+1}} \ldots e_{n_q}^{\varepsilon_{n_q}}$. Then $d\nu_{x_1} \ldots d\nu_{x_q} dg'$ will denote the following measure on $G^\Gamma$:

$$d\nu_{x_1}^{n_1}((g_1)^{\varepsilon_1},\ldots,(g_{n_1})^{\varepsilon_{n_1}}) \otimes \ldots \otimes d\nu_{x_q}^{n_q}((g_{n_{q-1}+1})^{\varepsilon_{n_{q-1}+1}},\ldots,(g_{n_q})^{\varepsilon_{n_q}}) \otimes dg'.$$

With this notation, we can set the following compact definitions:

$$Z(x_1,\ldots,x_q) = \int_{G^\Gamma} D \, d\nu_{x_1} \ldots d\nu_{x_q} dg', \tag{1.11}$$

$$dP_{(x_1,\ldots,x_q)} = \frac{1}{Z(x_1,\ldots,x_q)} D \, d\nu_{x_1} \ldots d\nu_{x_q} dg'. \tag{1.12}$$

The function $Z(x_1, \ldots, x_q)$ is called the *conditional partition function* with respect to $L_1, \ldots, L_q$. When $q = 0$, we agree to say that the conditional partition function is the constant $Z$ defined by (1.9). Observe that $Z(x_1, \ldots, x_q) > 0$ for all $x_1, \ldots, x_q \in G$. In order to check that $P_{(x_1, \ldots, x_q)}$ is the disintegration we are looking for, we must compute the distribution of $(h_{L_1}, \ldots, h_{L_q})$ under $P$. The following important result gives us in fact a bit more.

PROPOSITION 1.14. *Choose $0 < r \leq q$ and $x_{r+1}, \ldots, x_q \in G$. The distribution of $(h_{L_1}, \ldots, h_{L_r})$ under $P_{(x_{r+1}, \ldots, x_q)}$ is equal to*

$$\frac{Z(x_1, \ldots, x_q)}{Z(x_{r+1}, \ldots, x_q)} \, dx_1 \ldots dx_r,$$

*where each element $x_i$ refers to the loop $L_i$. In particular, the distribution of $(h_{L_1}, \ldots, h_{L_q})$ under $P$ is*

$$\frac{1}{Z} Z(x_1, \ldots, x_q) \, dx_1 \ldots dx_q.$$

PROOF. According to our convention about the partition function without argument, the last part of the statement is just the case $r = q$. Suppose $r$ given and choose a continuous function $f$ on $G^r$. In the computation, we will condition the measure successively with respect to $(h_{L_{r+1}}, \ldots, h_{L_q})$ and to $(h_{L_1}, \ldots, h_{L_r})$. The set of edges which do not appear in the decomposition of any of the loops $L_{r+1}, \ldots, L_q$ will be denoted by $\Gamma''$ and $dg''$ will denote the Haar measure on $G^{\Gamma''}$. We keep using the notation $dg'$ introduced above. We have:

$$\int_{G^\Gamma} f(h_{L_1}, \ldots, h_{L_r}) \, dP_{(x_{r+1}, \ldots, x_q)} =$$

$$= \frac{1}{Z(x_{r+1}, \ldots, x_q)} \int_{G^\Gamma} f(h_{L_1}, \ldots, h_{L_r}) D \, d\nu_{x_{r+1}} \ldots d\nu_{x_q} dg''$$

$$= \frac{1}{Z(x_{r+1}, \ldots, x_q)} \int_{G^r} \int_{G^\Gamma} f(h_{L_1}, \ldots, h_{L_r}) D \, d\nu_{x_1} \ldots d\nu_{x_q} dg' \, dx_1 \ldots dx_r$$

$$= \int_{G^r} f(x_1, \ldots, x_r) \left[ \frac{1}{Z(x_{r+1}, \ldots, x_q)} \int_{G^\Gamma} D \, d\nu_{x_1} \ldots d\nu_{x_q} dg' \right] dx_1 \ldots dx_r$$

$$= \int_{G^r} f(x_1, \ldots, x_r) \frac{Z(x_1, \ldots, x_q)}{Z(x_{r+1}, \ldots, x_q)} \, dx_1 \ldots dx_r. \qquad \square$$

COROLLARY 1.15. *The mapping $(x_1, \ldots, x_q) \mapsto P_{(x_1, \ldots, x_q)}$ is a disintegration of the measure $P$ with respect to the random variable $(h_{L_1}, \ldots, h_{L_q})$. This means that*

1. $(h_{L_1}, \ldots, h_{L_q}) = (x_1, \ldots, x_q)$ $P_{(x_1, \ldots, x_q)}$-*a.s.*
2. *If $\eta$ denotes the distribution of $(h_{L_1}, \ldots, h_{L_q})$ under $P$, we have*

$$P = \int_{G^q} P_{(x_1, \ldots, x_q)} \, d\eta(x_1, \ldots, x_q).$$

*The following property also holds:*

2'. *If $0 < r \leq q$ and $\eta_r$ denotes the distribution of $(h_{L_1}, \ldots, h_{L_r})$ under $P_{(x_{r+1}, \ldots, x_q)}$, then*

$$P_{(x_{r+1}, \ldots, x_q)} = \int_{G^r} P_{(x_1, \ldots, x_q)} \, d\eta_r(x_1, \ldots, x_r),$$

where $x_i$ always refers to the loop $L_i$.

PROOF. The first point is a direct consequence of the definition of $P_{(x_1,\ldots,x_q)}$. Property 2 is a consequence of 2' (take $r = q$), which results from Proposition 1.14 by an elementary computation. □

So far we have solved the problem of *deterministic* boundary conditions. Suppose now that we want to force the distribution of $(h_{L_1},\ldots,h_{L_q})$ to be equal to some prescribed probability measure on $G^q$, say $\beta$. Then we can simply put the measure $P_\beta$ on $G^\Gamma$, where

$$P_\beta = \int_{G^q} P_{(x_1,\ldots,x_q)}\, d\beta(x_1,\ldots,x_q). \tag{1.13}$$

This definition extends of course that of $P_{(x_1,\ldots,x_q)}$ in the sense that $P_{(x_1,\ldots,x_q)} = P_{\delta_{(x_1,\ldots,x_q)}}$.

REMARK 1.16. Beware the fact that, in general,

$$P_\beta \neq \frac{1}{K_\beta} \int_{G^q} D\, \nu_{x_1} \otimes \ldots \otimes \nu_{x_q} \otimes dg'\, d\beta(x_1,\ldots,x_q),$$

where

$$K_\beta = \int_{G^q} Z(x_1,\ldots,x_q)\, d\beta(x_1,\ldots,x_q).$$

REMARK 1.17. We are going to use such families of loops as $L_1,\ldots,L_q$ such a great number of times that we want to have a name for them : we shall call them *disintegration families*. What this term exactly means has been explained at the beginning of Section 1.5.2. Moreover, and unless explicitly mentioned, we will always have in mind that $q$ may be equal to 0: most statements about the conditional measure also hold for the free measure.

**1.5.3. Gauge transformations.** We have tuned the discrete Yang-Mills measure to be invariant under discrete gauge transformations, and it actually is, although this still has to be proved. But the invariance of the conditional versions of the measure is a different matter. For each $i = 1,\ldots,q$, let us denote by $m_i$ the base point of the loop $L_i$. Let us also introduce the notation $\text{Ad}(y)x = yxy^{-1}$ for the adjoint action of an element $y$ of $G$ on another element $x$ of $G$.

PROPOSITION 1.18. *Let $\phi = (\phi_v)_{v \in V(\Gamma)}$ be a discrete gauge transformation. For all $(x_1,\ldots,x_q) \in G^q$, the following equality holds:*

$$\phi_* P_{(x_1,\ldots,x_q)} = P_{(\text{Ad}(\phi_{m_1})x_1,\ldots,\text{Ad}(\phi_{m_q})x_q)}.$$

*In particular, if $q = 0$, $\phi_* P = P$.*

PROOF. For the sake of simplicity, we will write the proof in the case $q = 1$. Writing down the general case would not be more difficult but would require an unpleasant proliferation of notation.

Suppose that $L_1 = e_1 \ldots e_n$. Take a continuous function $f$ on $G^\Gamma$ and let us compute $(\phi_* P_{(x_1)})(f)$ using the left and right invariance of the Haar measure. It is equal to

$$\frac{1}{Z(x_1)} \int_{G^\Gamma} (f \circ \phi)(g) \prod_{F \in \mathcal{F}(\Gamma)} p_{\sigma(F)}(h_{\partial F}(g))\, d\nu_{x_1}^n(g_1,\ldots,g_n)\, dg_{n+1} \ldots dg_r$$

$$= \frac{1}{Z(x_1)} \int_{G^\Gamma} f(g) \prod_{F \in \mathcal{F}(\Gamma)} p_{\sigma(F)}((h_{\partial F} \circ \phi^{-1})(g))$$
$$d\nu^n_{x_1}(\phi_{e_1(1)}g_1(\phi_{e_1(0)})^{-1}, \ldots, \phi_{e_n(1)}g_n(\phi_{e_n(0)})^{-1}) \, dg_{n+1} \ldots dg_r$$
$$= \frac{1}{Z(x_1)} \int_{G^\Gamma} f(g) \prod_{F \in \mathcal{F}(\Gamma)} p_{\sigma(F)}(h_{\partial F}(g)) \, d\nu^n_{\phi_{e_n(1)}x_1(\phi_{e_1(0)})^{-1}}(g_1, \ldots, g_n)$$
$$dg_{n+1} \ldots dg_r$$
$$= \frac{Z(\mathrm{Ad}(\phi_{m_1})x_1)}{Z(x_1)} P_{(\mathrm{Ad}(\phi_{m_1})x_1)}(f),$$

with $m_1 = e_1(0) = e_n(1) = L_1(0)$. If $f$ is constant, we get the relation $Z(\mathrm{Ad}(\phi_{m_1})x_1) = Z(x_1)$ and the lemma is proved. □

The invariance property of the partition function that we have just proved deserves to be stated separately.

PROPOSITION 1.19. *For any $x_1, y_1, \ldots, x_q, y_q$ in $G$, the following equality holds:*
$$Z(\mathrm{Ad}(y_1)x_1, \ldots, \mathrm{Ad}(y_q)x_q) = Z(x_1, \ldots, x_q).$$

According to this result, the conditional partition function is really a function on $(G/\mathrm{Ad})^q$, where $G/\mathrm{Ad}$ is the quotient of $G$ by its own adjoint action.

We are now facing the following problem: Proposition 1.18 tells us that $P_{(x_1,\ldots,x_q)}$ is not gauge invariant in general, and we want to avoid as much as possible considering non gauge invariant objects. Fortunately, Proposition 1.20 below shows that some of the measures $P_\beta$ introduced at the end of Section 1.5.2 are gauge invariant.

We say that a measure $\beta$ on $G^q$ is *invariant* if it is invariant under the adjoint action of $G$ on each component of $G^q$. Properly speaking, we mean that for any continuous function $f$ on $G^q$ and any $y_1, \ldots, y_q \in G$, the following equality holds:
$$\int_{G^q} f(x_1, \ldots, x_q) \, d\beta(x_1, \ldots, x_q) = \int_{G^q} f(\mathrm{Ad}(y_1)x_1, \ldots, \mathrm{Ad}(y_q)x_q) \, d\beta(x_1, \ldots, x_q).$$

PROPOSITION 1.20. *If $\beta$ is invariant on $G^q$, then $P_\beta$ is gauge invariant.*

The proof is left to the reader as a simple consequence of Proposition 1.18.

Let us finish this paragraph by giving an important class of examples of invariant measures.

Let $\mathfrak{x}$ be an element of the quotient $G/\mathrm{Ad}$, that is, $\mathfrak{x}$ is a conjugacy class in $G^6$. Let $x$ be an element of this class. It follows from the right invariance of the Haar measure that the measure $\int_G \delta_{yxy^{-1}} \, dy$ on $G$ does not depend on the choice of $x$ in $\mathfrak{x}$. We shall denote it by $\delta_{\mathfrak{x}}$. Similarly, we will denote the measure $\delta_{\mathfrak{x}_1} \otimes \ldots \otimes \delta_{\mathfrak{x}_q}$ on $G^q$ by $\delta_{(\mathfrak{x}_1,\ldots,\mathfrak{x}_q)}$. It is obvious that this measure is invariant and we use the notation $P_{(\mathfrak{x}_1,\ldots,\mathfrak{x}_q)} = P_{\delta_{(\mathfrak{x}_1,\ldots,\mathfrak{x}_q)}}$ for the corresponding gauge invariant measure on $G^\Gamma$.

REMARK 1.21. Any invariant measure $\beta$ on $G^q$ induces a measure $\bar{\beta}$ on $(G/\mathrm{Ad})^q$ and it is readily checked that $P_\beta = \int_{(G/\mathrm{Ad})^q} P_{(\mathfrak{x}_1,\ldots,\mathfrak{x}_q)} \, d\bar{\beta}(\mathfrak{x}_1, \ldots, \mathfrak{x}_q)$.

---

[6]We will use gothic letters to denote conjugacy classes throughout this work.

## 1.6. Invariance under subdivision

Invariance under subdivision is by far the most important feature of the discrete theory. It allows one to prove that the distribution of the discrete holonomy along a given family of loops does not depend on the graph in which one computes it. This fact, apart from being satisfactory, makes it plausible that a "continuum limit" of the discrete measures can be defined as one considers increasingly fine sequences of graphs.

The fact that the heat kernel $(p_t)_{t>0}$ is a convolution semi-group will play a central role in the proof. This means that for any $x \in G$ and any $s, t$ such that $0 < s < t$,

$$\int_G p_s(xy^{-1}) p_{t-s}(y) \, dy = p_t(x).$$

In order to state the main result, consider two graphs $\Gamma_1 = \{e_1, \ldots, e_r\}$ and $\Gamma_2$ on $M$ and suppose that $\Gamma_2$ is finer than $\Gamma_1$. This means that each edge $e_i$ of $\Gamma_1$ is a path in $\Gamma_2^*$ and as such gives rise to a function $h_{e_i} : G^{\Gamma_2} \longrightarrow G$. These functions $h_{e_i}$ can be put together into a single function $(h_{e_1}, \ldots, h_{e_r}) : G^{\Gamma_2} \longrightarrow G^r = G^{\Gamma_1}$ which we will denote by $f_{\Gamma_1 \Gamma_2}$.

From now on, it will be sometimes necessary to specify the graph with respect to which we consider such objects as $P$, $Z$, or $D$.

THEOREM 1.22 (Invariance under subdivision). *Let $\Gamma_1$ and $\Gamma_2$ be two graphs on $M$ such that $\Gamma_2$ is finer than $\Gamma_1$. Let $L_1, \ldots, L_q$ be a disintegration family of loops of $\Gamma_1^*$ (see Remark 1.17). Let $x_1, \ldots, x_q$ be $q$ elements of $G$. Then*
1. *the map $f_{\Gamma_1 \Gamma_2} : G^{\Gamma_2} \longrightarrow G^{\Gamma_1}$ is surjective.*
2. *It satisfies $(f_{\Gamma_1 \Gamma_2})_* P^{\Gamma_2}_{(x_1, \ldots, x_q)} = P^{\Gamma_1}_{(x_1, \ldots, x_q)}$.*

Two lemmas are needed to prove this theorem. The first one shows that it is always possible to go from one graph to a finer graph by a finite sequence of elementary transformations.

LEMMA 1.23. *Let $\Gamma$ and $\Gamma'$ be two graphs such that $\Gamma \leq \Gamma'$. There exist an increasing sequence of graphs $\Gamma_0 = \Gamma \leq \Gamma_1 \leq \ldots \leq \Gamma_n \leq \ldots$, stationary of limit $\Gamma'$ and such that for any nonnegative $n$, one can transform $\Gamma_n$ into $\Gamma_{n+1}$ by one of the two following elementary operations:*
(V) *Add a vertex to $\Gamma_n$, i.e. replace an edge $e$ by two edges $e_1$ and $e_2$ such that $e = e_1 e_2$,*
(E) *Add an edge to $\Gamma_n$, provided at least one endpoint of this new edge is already a vertex of $\Gamma_n$.*

PROOF. We proceed by induction on $n$. $\Gamma_0$ is given, equal to $\Gamma$. Suppose $\Gamma_n$ has been constructed, with $\Gamma_n \leq \Gamma'$. Recall that $\mathcal{V}(\Gamma)$ is the set of vertices of $\Gamma$.
• We have $\mathcal{V}(\Gamma_n) \subset \mathcal{V}(\Gamma') \cap Supp(\Gamma_n)$. If this inclusion is a strict one, pick an element of $(\mathcal{V}(\Gamma') \cap Supp(\Gamma_n)) \setminus \mathcal{V}(\Gamma_n)$. It is a vertex of $\Gamma'$ which is on an edge of $\Gamma_n$ but is not one of its end points. By an operation (V), we add this vertex to $\Gamma_n$ and get $\Gamma_{n+1}$ such that $\Gamma'$ is still finer than $\Gamma_{n+1}$. Note that $\sharp \Gamma_{n+1} = \sharp \Gamma_n + 1$.
• If $\mathcal{V}(\Gamma_n) = \mathcal{V}(\Gamma') \cap Supp(\Gamma_n)$, then each edge of $\Gamma_n$ is an edge of $\Gamma'$. In other words, $\Gamma_n \subset \Gamma'$. If this inclusion is a strict one, there exists an edge of $\Gamma'$ which is not an edge of $\Gamma_n$ and by connectedness of $\Gamma'$ we may assume that this edge has at least one of its endpoints on $Supp(\Gamma_n)$. By an operation (E), we add this edge to $\Gamma_n$ and get $\Gamma_{n+1}$ which is still connected and such that $\Gamma_{n+1} \leq \Gamma'$. The relation

$\sharp\Gamma_{n+1} = \sharp\Gamma_n + 1$ is obvious.
- If $\Gamma_n = \Gamma'$, just set $\Gamma_{n+1} = \Gamma_n$.

At each step, the fact that $\Gamma_n$ is a graph implies that $\Gamma_{n+1}$ is also a graph : connectedness is preserved, as well as boundary properties. The faces of a graph are not modified by an operation (V) and it can happen that an operation (E) cuts a face into two pieces, or that it cuts it along a radial segment, and in both cases the resulting faces are still diffeomorphic to disks.

For each $n$, $\Gamma_n \leq \Gamma'$ implies $\sharp\Gamma_n \leq \sharp\Gamma'$ and we have noticed that elementary operations increase the cardinal of the graph. Thus, the sequence becomes stationary after a finite number of steps. □

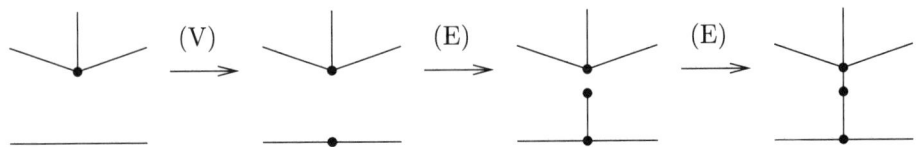

FIGURE 2. An example of elementary transformations.

The second lemma expresses a projectivity property.

LEMMA 1.24. *Let $\Gamma_1 \leq \Gamma_2 \leq \Gamma_3$ be three graphs. Then*
$$f_{\Gamma_1 \Gamma_3} = f_{\Gamma_1 \Gamma_2} \circ f_{\Gamma_2 \Gamma_3}.$$

PROOF. This follows from the associativity of the product in $G$. □

PROOF OF THEOREM 1.22. Lemmas 1.23 and 1.24 show that it is enough to prove the theorem when $\Gamma_2$ is deduced from $\Gamma_1$ by an elementary operation: the general case follows by composition.

We set $\beta = \delta_{(x_1,\ldots,x_q)}$ and use the notation $P_\beta = P_{(x_1,\ldots,x_q)}$ and $Z_\beta = Z(x_1,\ldots,x_q)$ to make expressions shorter.

1. If $\Gamma_2$ is deduced from $\Gamma_1$ by an operation (E), then $f_{\Gamma_1 \Gamma_2}$ is just the projection which forgets the factor associated with the new edge and it is of course surjective. In the case of an operation (V), $f_{\Gamma_1 \Gamma_2}$ preserves all factors except those associated with the two new edges, which get multiplied : it is also surjective.

2. Let us begin with the case of an operation of type (E). At this point, we need to introduce some notation. Set $\Gamma_1 = \{e_1,\ldots,e_r\}$, $\Gamma_2 = \{e_1,\ldots,e_r,e\}$ and call $F_0$ the face of $\Gamma_1$ which contains the new edge $e$. Two situations are possible : either $e$ has only one endpoint on $\partial F_0$ or it has both. In the first case, $F_0$ is still a face of $\Gamma_2$, but a new factor $ee^{-1}$ appears in the decomposition of its boundary. Let us consider the second case, where $e$ cuts $F_0$ into two faces $F_1$ and $F_2$. The boundaries of $F_0$, $F_1$ and $F_2$ can be written respectively as $\partial F_0 = c_1 c_2$, $\partial F_1 = c_1 e^{-1}$ and $\partial F_2 = ec_2$, where $c_1$ and $c_2$ belong to $\Gamma^*$. Take a continuous function $f$ on $G^{\Gamma_1}$. According to (1.12) (recall that $\beta = \delta_{(x_1,\ldots,x_q)}$), we have:

$$\int_{G^{\Gamma_1}} f\, d\left((f_{\Gamma_1 \Gamma_2})_* P_\beta^{\Gamma_2}\right) = \frac{1}{Z_\beta^{\Gamma_2}} \int_{G^{\Gamma_2}} f(g_1,\ldots,g_r) p_{\sigma(F_1)}(g_{r+1}^{-1} h_{c_1}) p_{\sigma(F_2)}(h_{c_2} g_{r+1})$$

$$\prod_{F\in\mathcal{F}(\Gamma_1)\setminus F_0} p_{\sigma(F)}(h_{\partial F})\,d\nu_{x_1}\ldots d\nu_{x_q}dg',$$

where $g_{r+1}$ is the component associated with $e$. Since the $L_i$'s are paths in $\Gamma_1$, the new edge $e$ is not involved in their decomposition. Thus we can isolate $dg_{r+1}$ in $dg'$ and integrate against it. By using the convolution property of the heat kernel we find

$$\frac{1}{Z_\beta^{\Gamma_2}}\int_{G^{\Gamma_1}} f(g_1,\ldots,g_r)p_{\sigma(F_1)+\sigma(F_2)}(h_{c_2}h_{c_1})\prod_{F\in\mathcal{F}(\Gamma_1)\setminus F_0} p_{\sigma(F)}(h_{\partial F})\,d\nu_{x_1}\ldots d\nu_{x_q}dg'$$

$$=\frac{Z_\beta^{\Gamma_1}}{Z_\beta^{\Gamma_2}}\int_{G^{\Gamma_1}} f\,dP_\beta^{\Gamma_1}.$$

If $f$ is constant, we get $Z_\beta^{\Gamma_1}=Z_\beta^{\Gamma_2}$ and the result follows.

The first case, where the new edge does not cut $F_0$ into two faces is in fact simpler: the factor $ee^{-1}$ which has appeared in $\partial F_0$ vanishes in all computations, because $f$ does not depend on the holonomy along $e$.

Let us consider now the case of an operation (V). This is in principle the simplest case, but the fact that the new vertex can be on the image of one of the $L_i$'s makes it a bit longer to write down.

Set $\Gamma_1=\{e,e_3,\ldots,e_r\}$ and $\Gamma_2=\{e_1,e_2,e_3\ldots,e_r\}$, with $e=e_1e_2$. Note that $e$ can be on the boundary of one or two faces, but that this does not matter really for the proof. Given a continuous function $f$ on $G^{\Gamma_1}$ as above, we get:

$$\int_{G^{\Gamma_1}} f\,d\left((f_{\Gamma_1\Gamma_2})_* P_\beta^{\Gamma_2}\right)=\frac{1}{Z_\beta^{\Gamma_2}}\int_{G^{\Gamma_2}} f(g_2g_1,g_3,\ldots,g_r)D^{\Gamma_1}(g_2g_1,g_3,\ldots,g_r)$$

$$d\nu_{x_1}\ldots d\nu_{x_q}dg',$$

where $(g_1,g_2,g_3,\ldots,g_r)$ denotes the generic element of $G^{\Gamma_2}$. If $e$ is not involved in the decomposition of any of the $L_i$'s, we can isolate $dg_1dg_2$ in $dg'$ and, by integrating against the left-invariant measure $dg_1$, the dependence in $g_2$ disappears:

$$=\frac{1}{Z_\beta^{\Gamma_2}}\int_{G^{\Gamma_1}} f(g,g_3,\ldots,g_r)D^{\Gamma_1}(g,g_3,\ldots,g_r)\,d\nu_{x_1}\ldots d\nu_{x_q}dg'=\frac{Z_\beta^{\Gamma_1}}{Z_\beta^{\Gamma_2}}\int_{G^{\Gamma_1}} f\,dP_\beta^{\Gamma_1}.$$

The conclusion follows as before. If $e$ appears in the decomposition of, say, $L_1$, we may assume that $L_1=e_1e_2\ldots e_m$, with $m\geq 3$. We can write $d\nu_{x_1}$ in a convenient way, putting the condition on $g_m$, which is necessarily distinct from $g_1$. We get:

$$\frac{1}{Z_\beta^{\Gamma_2}}\int_{G^{\Gamma_1}} f(g_2g_1,g_3,\ldots,\widetilde{g_m},\ldots,g_r)D^{\Gamma_1}(g_2g_1,g_3,\ldots,\widetilde{g_m},\ldots,g_r)$$

$$dg_1dg_2dg_3\ldots dg_{m-1}d\nu_{x_2}\ldots d\nu_{x_q}dg',$$

where $\widetilde{g_m}=x_1(g_{m-1}\ldots g_3g_2g_1)^{-1}$. This is equal to

$$\frac{1}{Z_\beta^{\Gamma_2}}\int_{G^{\Gamma_1}} f(g,g_3,\ldots,\widetilde{g_m},\ldots,g_r)D^{\Gamma_1}(g,g_3,\ldots,\widetilde{g_m},\ldots,g_r)$$

$$dgdg_3\ldots dg_{m-1}d\nu_{x_2}\ldots d\nu_{x_q}dg',$$

with $\widetilde{g_m}$ being now equal to $x_1(g_{m-1}\ldots g_3g)^{-1}$. The expression becomes:

$$\frac{Z_\beta^{\Gamma_1}}{Z_\beta^{\Gamma_2}}\int_{G^{\Gamma_1}} f\,dP_\beta^{\Gamma_1}$$

and the result is proved. □

COROLLARY 1.25. *Let $\Gamma_1$ be a graph and $c_1,\ldots,c_n$ a family of paths of $\Gamma_1^*$. Then for any graph $\Gamma_2$ finer than $\Gamma_1$, $c_1,\ldots,c_n$ are also paths of $\Gamma_2^*$ and the distribution of the discrete holonomy along $c_1,\ldots,c_n$ is the same in $\Gamma_1$ and $\Gamma_2$. More precisely, the distribution of $(h_{c_1}^{\Gamma_1},\ldots,h_{c_n}^{\Gamma_1})$ on $(G^{\Gamma_1},P_\beta^{\Gamma_1})$ and that of $(h_{c_1}^{\Gamma_2},\ldots,h_{c_n}^{\Gamma_2})$ on $(G^{\Gamma_2},P_\beta^{\Gamma_2})$ are equal.*

PROOF. There remains only to check that $(h_{c_1}^{\Gamma_1},\ldots,h_{c_n}^{\Gamma_1})\circ f_{\Gamma_1\Gamma_2} = (h_{c_1}^{\Gamma_2},\ldots,h_{c_n}^{\Gamma_2})$. This is true if the $c_i$'s are edges of $\Gamma_1$, and the general case follows by multiplicativity. □

A very important property of the conditional partition functions has emerged during the proof of Theorem 1.22.

PROPOSITION 1.26. *Let $\Gamma_1$ and $\Gamma_2$ be two graphs such that $\Gamma_1 \leq \Gamma_2$. Take a disintegration family $L_1,\ldots,L_q$ in $\Gamma_1^*$ and $x_1,\ldots,x_q$ in $G$. Then*
$$Z^{\Gamma_1}(x_1,\ldots,x_q) = Z^{\Gamma_2}(x_1,\ldots,x_q).$$

Let us discuss briefly the choice of the heat kernel in the definition of $P$, which is the key of the physical relevance of the theory. A physicist, A. Migdal [**32**], although he did not name it explicitly, was the first to derive from physical considerations the fact that the heat kernel was the correct central function to be used in the definition of $P$. Nevertheless, this choice is certainly not the unique possible one, nor perhaps the unique sensible one. It is possible to construct a discrete theory using any other convolution semigroup with enough invariance properties. For example, S. Albeverio, R. Høegh Krohn and H. Holden investigated some properties of the random fields one could obtain this way in [**2**].

## 1.7. Invariance under area-preserving diffeomorphisms

Apart from the differentiable structure, the only geometrical structure on $M$ is the Lebesguian surface measure $\sigma$. We are going to show that the discrete theory is invariant under *area-preserving diffeomorphisms*, those which preserve this structure.

Let $\psi : M \longrightarrow M$ a diffeomorphism such that $\psi_*\sigma = \sigma$ and let $\Gamma$ be a graph. Then $\psi$ transforms $\Gamma$ edge by edge into another graph $\psi(\Gamma)$ and induces a bijection which preserves the area between the set of faces of $\Gamma$ and that of $\psi(\Gamma)$. This makes it clear that the natural bijection which $\psi$ also induces between $G^\Gamma$ and $G^{\psi(\Gamma)}$ preserves the discrete Yang-Mills measure. Let us state this invariance property in a slightly more general context.

PROPOSITION 1.27. *Let $(M_1,\sigma_1)$ and $(M_2,\sigma_2)$ be two surfaces endowed with Lebesguian measures and let $\psi : M_1 \longrightarrow M_2$ be a diffeomorphism such that $\psi_*\sigma_1 = \sigma_2$. Let $\Gamma_1$ be a graph on $M_1$ and $\Gamma_2 = \psi(\Gamma_1)$ the corresponding graph on $M_2$. Finally, let $L_1,\ldots,L_q$ be a disintegration family on $M_1$ and $\psi(L_1),\ldots,\psi(L_q)$ the corresponding family on $M_2$. Still denoting by $\psi : G^{\Gamma_1} \longrightarrow G^{\Gamma_2}$ the induced bijection, we have*
$$\psi_* P^{\Gamma_1}_{(x_1,\ldots,x_q)} = P^{\Gamma_2}_{(x_1,\ldots,x_q)}.$$

In particular, given a family $(c_1, \ldots, c_n)$ of paths in $\Gamma_1^*$, this shows that the discrete holonomy along $(c_1, \ldots, c_n)$ and the discrete holonomy along $(\psi(c_1), \ldots, \psi(c_n))$ have the same distribution.

## 1.8. Two examples

In this section, we consider two basic situations. First, we consider an open path and show that the distribution of the holonomy along this path is uniform, whatever this path looks like. This provides an example of the power of gauge invariance. Then, we give a simple but important estimate for the mean of the distance to the unit element of the holonomy along a small simple loop.

### 1.8.1. Holonomy along an open path.
Let $\Gamma$ be a graph and $L_1, \ldots, L_q$ a disintegration family of $\Gamma^*$ (see Remark 1.17). Pick $q$ conjugacy classes $\mathfrak{x}_1, \ldots, \mathfrak{x}_q$ in $G/\mathrm{Ad}$ and recall the probability measure $P_{(\mathfrak{x}_1, \ldots, \mathfrak{x}_q)}$ on $G^\Gamma$ defined in Sections 1.5.2 and 1.5.3.

PROPOSITION 1.28. *If $c$ is a path of $\Gamma^*$ with distinct endpoints, then the distribution of $h_c$ under $P_{(\mathfrak{x}_1, \ldots, \mathfrak{x}_q)}$ is the uniform distribution on $G$, that is, the normalized Haar measure.*

PROOF. The proof is simple and relies upon the fact that $P_{(\mathfrak{x}_1, \ldots, \mathfrak{x}_q)}$ is gauge invariant, according to Proposition 1.20.

Let $f$ be a continuous function on $G$ and $\phi$ a discrete gauge transformation. We have

$$\int_{G^\Gamma} f(h_c) \, dP_{(\mathfrak{x}_1, \ldots, \mathfrak{x}_q)} = \int_{G^\Gamma} f(h_c \circ \phi) \, dP_{(\mathfrak{x}_1, \ldots, \mathfrak{x}_q)}$$
$$= \int_{G^\Gamma} f((\phi_{c(1)})^{-1} h_c \phi_{c(0)}) \, dP_{(\mathfrak{x}_1, \ldots, \mathfrak{x}_q)}.$$

Since $c(0) \neq c(1)$, $\phi_{c(0)}$ and $\phi_{c(1)}$ are arbitrary elements of $G$, so that the distribution of $h_c$ is right and left invariant: it is the Haar measure. □

This result is a kind of stochastic counterpart to the geometric fact that, given a connection, there is no intrinsic way to attach an element of $G$ to an open path. This can help to justify the fact that we shall in the end restrict ourselves to loops in the definition of the continuous Yang-Mills measure.

### 1.8.2. Holonomy around a small disk.
Let $\rho$ be the function defined on $G$ by $\rho(x) = d(1, x)$, where $d$ is the bi-invariant Riemannian distance with unit total volume and 1 is the unit element. The following result will play an essential role in the construction of the continuous measure.

PROPOSITION 1.29. *Let $L_1, \ldots, L_q$ be a disintegration family on $M$. There exist two positive constants $s$ and $C$ such that, for any graph $\Gamma$ such that $L_1, \ldots, L_q \in \Gamma^*$, any simple loop $l \in \Gamma^*$ which bound a closed disk $D$ such that $\sigma(D) \leq s$ and no $L_i$ is fully contained in $D$, and any $x = (x_1, \ldots, x_q) \in G^q$, one has*

$$\int_{G^\Gamma} \rho(h_l) \, dP_{(x_1, \ldots, x_q)} \leq C \sqrt{\sigma(D)}. \tag{1.14}$$

We are going to estimate the real-valued random variable $\rho(h_l)$ using a standard short-time estimate for the heat kernel on $G$. Let $d$ denote the dimension of $G$.

PROPOSITION 1.30 ([43], Th. V.4.3). *There exists a positive constant $C$ such that for all $t \in (0,1)$, all $x \in G$,*

$$\frac{1}{C} t^{-\frac{d}{2}} e^{-\frac{C\rho(x)^2}{t}} \leq p_t(x) \leq C t^{-\frac{d}{2}} e^{-\frac{\rho(x)^2}{Ct}}.$$

The majoration in this inequality allows us to prove the following estimate.

PROPOSITION 1.31. *For any $q \geq 1$, it is the case that*

$$\int_G \rho(x)^q p_t(x)\, dx = O(t^{\frac{q}{2}})$$

*for $t$ in a neighbourhood of zero.*

PROOF. Fix $q \geq 1$. For any radius $R > 0$, the majoration in Proposition 1.30 implies

$$\int_{\rho \geq R} \rho(x)^q p_t(x)\, dx \leq C t^{-\frac{d}{2}} e^{-\frac{R^2}{Ct}} \operatorname{diam}(G)^q = o(t^n) \quad \forall n \geq 0. \tag{1.15}$$

In a neighbourhood of the identity, we need to be more careful. Let us denote by $\mathfrak{g}$ the Lie algebra of $G$ and choose $R_0$ small enough for the exponential map $\exp : \mathfrak{g} \longrightarrow G$ to be a diffeomorphism from the ball[7] $B(0, R_0) \subset \mathfrak{g}$ onto the set $\{\rho < R_0\} \subset G$. A fundamental property of the exponential map is that it preserves radial distances. Using polar coordinates $(r, \theta)$ on $B(0, R_0)$, where $\theta$ belongs to the unit sphere $S^{d-1}$ and $r > 0$, this means that $\rho(\exp(r, \theta)) = r$. Moreover, the Haar measure on $G$ is associated with a non-vanishing smooth volume form, so that its image by $\exp^{-1}$ is a measure on $B(0, R_0)$ which is equivalent to the Lebesgue measure $r^{d-1}\, dr d\theta$ with a smooth density bounded above and below. The majoration of Proposition 1.30 now implies

$$\begin{aligned}
\int_{\rho < R_0} \rho(x)^q p_t(x)\, dx &\leq C_1 \int_{[0,R_0) \times S^{d-1}} r^q\, p_t(\exp(r,\theta)) r^{d-1}\, dr d\theta \\
&\leq C_2\, t^{-\frac{d}{2}} \int_0^{R_0} r^{d-1+q} e^{-\frac{r^2}{Ct}}\, dr \\
&\leq C_2\, t^{\frac{q}{2}+\frac{d}{2}} t^{-\frac{d}{2}} \int_0^\infty \left(\frac{r}{\sqrt{t}}\right)^{d-1+q} e^{-\frac{1}{C}\left(\frac{r}{\sqrt{t}}\right)^2} d\left(\frac{r}{\sqrt{t}}\right) \\
&\leq C_3\, t^{\frac{q}{2}}. \tag{1.16}
\end{aligned}$$

Combining (1.15) and (1.16), we see that for $t$ in a neighbourhood of 0, we have

$$\begin{aligned}
\int_G \rho(x)^q p_t(x)\, dx &= \int_{\rho < R_0} \rho(x)^q p_t(x)\, dx + \int_{\rho \geq R_0} \rho(x)^q p_t(x)\, dx \\
&\leq C_3\, t^{\frac{q}{2}} + o(t^{\frac{q}{2}}) = O(t^{\frac{q}{2}}),
\end{aligned}$$

and the result is proved. □

In order to understand the content of Proposition 1.29, suppose for one second that $q = 0$ and that $M$ is a sphere. Then the set of edges of $l$ forms a graph on

---

[7] Recall that the bi-invariant Riemannian metric on $G$ endows $\mathfrak{g} = T_1 G$ with a scalar product invariant by adjunction. If $G = SU(2)$, then $\mathfrak{g} = \mathfrak{su}(2)$ is the space of traceless anti-hermitian $2 \times 2$ matrices and, up to a constant, the scalar product is the Killing form $K(A,B) = -\operatorname{tr}(A^* B)$.

$M$ and the distribution of $h_l$ under $P$, computed in this graph, which is legitimate according to Corollary 1.25, is given by

$$\int_{G^\Gamma} f(h_l)\, dP = \frac{1}{p_{\sigma(M)}(1)} \int_G f(x) p_{\sigma(D)}(x) p_{\sigma(M-D)}(x^{-1})\, dx, \qquad (1.17)$$

where $f$ is any continuous function on $G$. Given the estimation of Proposition 1.31, this makes Proposition 1.29 almost obvious in this case. Incidentally, (1.17) shows that $h_l$ has the distribution of a Brownian bridge on $G$ starting at 1, of length $\sigma(M)$, at time $\sigma(D)$.

The complete proof of Proposition 1.29 is now a bit longer than one would expect after this remark, but it is really no more than a technical variation on the same idea, with the complication due to the presence of the disintegration family and to the fact that one has to choose a convenient graph in which to perform the computation.

PROOF OF PROPOSITION 1.29. We want to compute the distribution of $h_l$ under $P_{(x_1,\ldots,x_q)}$ and, according to the subdivision invariance property, we can do this in any graph $\Gamma_1$ such that $L_1, \ldots, L_q$ and $l$ belong to $\Gamma_1^*$. We are going to construct carefully such a graph.

We start by taking the edges of $\Gamma$ which appear in the decomposition of the $L_i$'s. In fact, since we have made the assumption that none of the $L_i$'s is fully contained in $\overline{D}$, we may now assume that each $L_i$ has even an edge outside $\overline{D}$. This may require to subdivide some edges of $\Gamma$ and so we do if necessary. We get in this way a pregraph $\Gamma_L$ on $M$ and we denote by $S$ the infimum of the areas of the faces of $\Gamma_L$. Now let us add to this pregraph the edges of $l$. We still get a pregraph, each face of which is diffeomorphic to the interior of a surface with boundary. On each of these faces, Proposition 1.8 tells us that we can extend our pregraph to a graph with the same number of faces, that is, with only one face. Putting all these graphs together, we finally get a graph $\Gamma_1$. The computation in $\Gamma_1$ reads:

$$\int_{G^{\Gamma_1}} \rho(h_l)\, dP_{(x_1,\ldots,x_q)} = \frac{1}{Z(x_1,\ldots,x_q)} \int_{G^{\Gamma_1}} \rho(h_l) \prod_{F \in \mathcal{F}(\Gamma_1)} p_{\sigma(F)}(h_{\partial F})\, d\nu_{x_1} \ldots d\nu_{x_q} dg'.$$

We are going to distinguish between the faces which are inside $D$ and those which are not. Take a face $F$ outside $D$. It is contained in one of the faces, say $F_0$, of $\Gamma_L$. In fact, up to a set of measure 0, we have $F = F_0 - (F_0 \cap D)$. Recall that we have defined $S$ such that $\sigma(F_0) \geq S$: if we assume now that $\sigma(D) \leq S/2$, then we get $\sigma(F) \geq S/2$. For all $0 < u < t$, the computation

$$p_t(x) = \int_G p_u(xy^{-1}) p_{t-u}(y)\, dy \leq \| p_u \|_\infty \int_G p_{t-u}(y)\, dy \leq \| p_u \|_\infty$$

shows that $t \mapsto \| p_t \|_\infty$ is a non-increasing function on $(0,\infty)$ and indeed

$$\int_{G^{\Gamma_1}} \rho(h_l)\, dP_{(x_1,\ldots,x_q)} \leq \frac{1}{Z(x_1,\ldots,x_q)} \int_{G^{\Gamma_1}} \rho(h_l) \prod_{F \not\subset D} \| p_{S/2} \|_\infty$$
$$\prod_{F \subset D} p_{\sigma(F)}(h_{\partial F})\, d\nu_{x_1} \ldots d\nu_{x_q} dg'.$$

The integrand depends now only on the edges of $\Gamma_1$ which are inside $D$. The assumption that each $L_i$ has at least one edge outside $D$ allows us to put the condition in $d\nu_{x_i}$ on such an edge, so that this condition disappears. We get

$$\int_{G^{\Gamma_1}} \rho(h_l) \, dP_{(x_1,\ldots,x_q)} \leq \frac{\prod_{F \not\subset D} \|p_{S/2}\|_\infty}{Z(x_1,\ldots,x_q)} \int_{G^{\Gamma_1 \cap D}} \rho(h_l) \prod_{F \subset D} p_{\sigma(F)}(h_{\partial F}) dg,$$

where $\Gamma_1 \cap D$ denotes the restriction of $\Gamma_1$ to $D$. We are now working on $\overline{D}$ instead of $M$ and the subdivision invariance on $\overline{D}$ allows us to replace $\Gamma_1 \cap D$ by the graph which contains only the edges of $l$. The unique face of this graph is $D$ itself and we get the inequality

$$\int_{G^{\Gamma_1}} \rho(h_l) \, dP_{(x_1,\ldots,x_q)} \leq \frac{\prod_{F \not\subset D} \|p_{S/2}\|_\infty}{Z(x_1,\ldots,x_q)} \int_G \rho(x) p_{\sigma(D)}(x) \, dx.$$

Combined with the result of Proposition 1.31, this establishes (1.14) for $\sigma(D)$ small enough, provided we show that $Z(x_1,\ldots,x_q)$ is bounded below uniformly in $x_1,\ldots,x_q$.

The key to this fact is the minoration in Proposition 1.30, which shows that $p_t$ is bounded below by a positive constant for each $t > 0$. Now, the integrand in the expression

$$Z(x_1,\ldots,x_q) = \int_{G^{\Gamma_1}} \prod_{F \in \mathcal{F}(\Gamma_1)} p_{\sigma(F)}(h_{\partial F}) \, d\nu_{x_1} \ldots d\nu_{x_q} dg'$$

is bounded below by a positive constant and the result follows. $\square$

## 1.9. Discrete Abelian theory

The case of an Abelian structure group $G$ deserves a special treatment for several reasons. The most important one is perhaps that $G = U(1)$ corresponds to the case of quantum electrodynamics, which has been the first (extremely) successful gauge theory. A second reason is that it is a relatively simple and concrete example, insofar
– the geometry involved is slightly less complicated than in the general case,
– we have an explicit expansion for the heat kernel,
– the combinatorics of the discrete theory is much easier.
Finally, it is the situation where the heuristic link between the Yang-Mills measure and a Gaussian measure on the space of square-integrable $\mathfrak{g}$-valued functions on $M$, as explained in the introduction, is the most direct. To explore this link in the light of our construction of the continuous Yang-Mills measure and to discuss to what extent it is still valid without the commutativity assumption on $G$ is the main purpose of Chapters 3 and 4. The present section is indeed the discrete prelude to the continuous study presented in Chapter 3.

Apart from the the chapters of [38] A. Sengupta has devoted to it, the Abelian case has rarely been studied separately. An exception is the paper [11], in which C. Becker has been the first to analyze the Abelian theory on the complex plane by using the winding numbers of loops. Our extensive use of the double-layer potential is based on the same idea.

This section is written in the case $G = U(1)$, whereas, according to our assumptions on $G$, the most general Abelian case would be $G = U(1)^n$. As a matter of fact, the choice of $G = U(1)$ makes a lot of things easier to write down, and does not make the mathematics any poorer.

### 1.9.1. Decomposition of cycles: fundamental systems.
Throughout this section, we fix a graph $\Gamma$ on $M$. Our initial purpose is to analyze in detail the distribution of the family of random variables $(h_c)_{c \in \Gamma^*}$ under the discrete Yang-Mills measure.

I say *initial* purpose, because the first implication of $G$ being Abelian is that we want in fact to study a slightly different family of random variables. Indeed, the function $h_c : G^\Gamma \longrightarrow G$ associated with a path $c \in \Gamma^*$ depends only on the *number of occurences* of each edge of $\Gamma$ in the decomposition of $c$, not on the *order* in which these edges appear. Let us put $\Gamma = \{e_1, \ldots, e_r\}$. In more formal terms, we can say that the function $h_c$ depends only on the image of $c$ by the natural morphism of monoids $\Gamma^* \longrightarrow \mathbb{Z}^\Gamma$ which sends $e_i$ to $(0, \ldots, 1, \ldots, 0)$ with a 1 at the $i$-th place. Conversely, each element of $\mathbb{Z}^\Gamma$ determines without ambiguity a function from $U(1)^\Gamma$ into $U(1)$, though not necessarily of the form $h_c$ for any $c \in \Gamma^*$. This allows us in particular to consider *linear combinations* of paths. Actually, we are chiefly interested in loops, and we denote by $C\Gamma \subset \mathbb{Z}^\Gamma$ the set of linear combination of loops, also called *cycles*. It seems now natural to consider the family of random variables $(h_c)_{c \in C\Gamma}$ instead of $(h_c)_{c \in \Gamma^*}$. Note however that $\Gamma^*$ cannot be imbedded into $C\Gamma$ because $\Gamma^*$ contains open paths and note also that elements of $C\Gamma$ do not have an origin. In an Abelian context, this is not a big loss, for two loops which differ only by the choice of their origin have the same holonomy.

We gain a $\mathbb{Z}$-module structure by looking at this larger family and we want to prove a decomposition result in order to express the distribution of the whole family of cycles as a function of the distribution of a finite generating subfamily. This involves the topology of $M$, so let us start by recalling a classical result.

THEOREM 1.32. *Let $g$ be the genus of $M$ and $p$ the number of connected components of $\partial M$. Then*

$$H_1(M; \mathbb{Z}) \simeq \begin{cases} \mathbb{Z}^{2g} & \text{if } p = 0 \\ \mathbb{Z}^{2g+p-1} & \text{if } p > 0. \end{cases}$$

If $M$ is closed, that is, if $p = 0$, one can construct a system of representatives for a basis of $H_1(M)$ by taking $2g$ loops according to the usual picture, two around each handle of $M$ intersecting only once and transversally (see [**14**] for example). If $p > 0$, then one can start by taking $2g$ loops of $M$ generating the $H_1$ of a minimal closure of $M$ (see Definition 1.1) and then add any $p-1$ components of $\partial M$.

Let us fix such a system of representatives $\ell_1, \ldots, \ell_{2g}$ in $\Gamma^*$ and $p-1$ loops $N_1, \ldots, N_{p-1}$ which we denote just as the corresponding boundary components. The $\ell_i$'s can be obtained as paths of $\Gamma^*$ by deforming an arbitrary system of generators using the same technique as in the proof of Proposition 1.4. Observe that we do not put any restriction on the way the $\ell'_i s$ intersect each other. By definition of a basis of $H_1(M; \mathbb{Z})$, any cycle $c$ in $C\Gamma$ has an unique decomposition

$$c = \lambda_1 \ell_1 + \ldots + \lambda_{2g} \ell_{2g} + \nu_1 N_1 + \ldots + \nu_{p-1} N_{p-1} + c^\perp,$$

where $\lambda_1, \ldots, \lambda_{2g}, \nu_1, \ldots, \nu_{p-1}$ are integers and $c^\perp \in C\Gamma$ is a cycle homologous to zero. Let us denote by $C_0\Gamma$ the submodule of $C\Gamma$ spanned by the cycles homologous to zero. We need now to understand what the cycles of $C_0\Gamma$ look like.

PROPOSITION 1.33. *If $\partial M$ is empty (resp. non-empty), the boundaries of all faces except one chosen arbitrarily (resp. of all faces) form a basis of the submodule $C_0\Gamma$ of $C\Gamma$.*

Let us admit this proposition for a short while. Let $\mathcal{F}(\Gamma) = \{F_1, \ldots, F_n\}$ be the set of faces of $\Gamma$ and let us choose for each $F_i$ a cycle $\partial F_i$ whose image is the boundary of $F_i$. Note that we are not making any assumption on the orientation of the cycles $\partial F_i$. We can decompose $c^\perp$ further and write, for some integers $\mu_1, \ldots, \mu_n$ :

$$c = \lambda_1 \ell_1 + \ldots + \lambda_{2g} \ell_{2g} + \nu_1 N_1 + \ldots + \nu_{p-1} N_{p-1} + \mu_1 \partial F_1 + \ldots + \mu_n \partial F_n, \quad (1.18)$$

the decomposition being non unique if $M$ is closed. The relation (1.18), together with the multiplicativity of the holonomy, allows us to state the following lemma.

LEMMA 1.34. *The distribution of the family $(h_c)_{c \in C\Gamma}$ is completely determined by that of the following finite collection of random variables:*

$$(h_{\ell_1}, \ldots, h_{\ell_{2g}}, h_{N_1}, \ldots, h_{N_{p-1}}, h_{\partial F_1}, \ldots, h_{\partial F_n}). \quad (1.19)$$

*We will call such a finite collection a* fundamental system.

PROOF OF PROPOSITION 1.33. To begin with, suppose that $M$ has no boundary. We proceed by induction on $n = \sharp \mathcal{F}(\Gamma)$. If $\Gamma$ has only one face, the result is clearly true. Suppose that it is true for a graph with $n-1$ faces and let $\Gamma = \{e_1, \ldots, e_r\}$ be a graph with $n$ faces. There is an edge of $\Gamma$, say $e_r$, which is on the boundary of two distinct faces, say $F_{n-1}$ and $F_n$. Let $\Gamma' = \{e_1, \ldots, e_{r-1}\}$ be the graph obtained by removing $e_r$. It has $n-1$ faces $F_1, \ldots, F_{n-2}, F_{n-1} \cup F_n$, up to negligible sets. Let $c$ be a cycle of $C_0\Gamma$. We can decompose it uniquely in $c = c' + pe_r$ with $p \in \mathbb{Z}$ and $c' \in \mathbb{Z}^{\Gamma'}$. We can also write $\partial F_{n-1} = e_r + f$ with $f \in \mathbb{Z}^{\Gamma'}$. So, we have $c = (c' - pf) + p\partial F_{n-1}$. This shows that $c' - pf$ belongs not only to $\mathbb{Z}^{\Gamma'}$ but to $C\Gamma'$, actually to $C_0\Gamma'$. By induction, $c' - pf$, which is homologous to zero in $\Gamma'$, is a linear combination of $\partial F_1, \ldots, \partial F_{n-2}$. Thus, $\partial F_1, \ldots, \partial F_{n-1}$ generate the submodule of homologically trivial cycles in $C\Gamma$. On the other hand, $\partial F_1, \ldots, \partial F_{n-2}$ are linearly independent by induction and $\partial F_{n-1}$ is independent of the submodule that they generate, because it contains the edge $e_r$. This gives the result when $M$ is closed.

If $M$ has a boundary, consider a minimal closure $i_1 : M \longrightarrow M_1$ of $M$ and identify $M$ with $i_1(M)$. Let $c$ be a cycle homologous to zero in $M$. It is also homologous to zero in $M_1$ and can be decomposed using the result on $M_1$ into :

$$c = \sum_{F_i \in \mathcal{F}(\Gamma), F_i \subset M} \mu_i \partial F_i + \nu_1 N_1 + \ldots + \nu_{p-1} N_{p-1},$$

because the $N_i$'s are the boundaries of the faces of $\Gamma$ on $M_1 - M$. This decomposition gives, in $H_1(M)$,

$$[c] = 0 = \nu_1 [N_1] + \ldots + \nu_{p-1} [N_{p-1}],$$

which implies $\nu_1 = \ldots = \nu_{p-1} = 0$ and indeed $c = \mu_1 \partial F_1 + \ldots + \mu_n \partial F_n$. The independence of the $\partial F_i$'s on $M_1$ implies their independence on $M$. □

### 1.9.2. Study of a fundamental system.

We are going to study the distribution of the fundamental system (1.19) under the Yang-Mills measure conditioned by its holonomy along the boundary components of $M$. This discussion could easily be extended to the case of arbitrary boundary conditions, by using the relation

(1.13).

Consider the disintegration family $N_1, \ldots, N_p$, take an element $(x_1, \ldots, x_p)$ of $U(1)^p$ and let $P_{(x_1,\ldots,x_p)}$ be the corresponding conditional measure. Part of the problem is trivial, since by definition, under $P_{(x_1,\ldots,x_p)}$, $(h_{N_1}, \ldots, h_{N_{p-1}}) = (x_1, \ldots, x_{p-1})$ almost surely. The next result solves the other part of the problem.

PROPOSITION 1.35. *Under the measure $\nu_{x_1} \otimes \ldots \otimes \nu_{x_p} \otimes dg'$ on $U(1)^\Gamma$, the variables $h_{\ell_1}, \ldots, h_{\ell_{2g}}, h_{\partial F_1}, \ldots, h_{\partial F_{n-1}}$ are uniform and independent on $U(1)$.*

PROOF. We prove this by imbedding $U(1)$ in $\mathbb{C}$, as the subgroup of complex numbers of modulus 1, and by computing the joint characteristic function of $(h_{\ell_1}, \ldots, h_{\ell_{2g}}, h_{\partial F_1}, \ldots, h_{\partial F_{n-1}})$ seen as a $\mathbb{C}^{2g+n-1}$-valued random variable. Note that, since this distribution is supported by $\{z : |z| = 1\}^{2g+n-1}$, it is characterized by the values of its characteristic function at all $(2g+n-1)$-tuples of integers.

In order to simplify the notation, we choose an orientation of $M$ and assume that each $N_i \subset \partial M$ and each $\partial F_i$ is oriented according to the usual convention. Note that, with this convention, the equality $\sum_{j=1}^n \partial F_j = \sum_{i=1}^p N_i$ holds in $\mathbb{Z}^\Gamma$.

Let $\lambda = (\lambda_1, \ldots, \lambda_{2g})$ and $\mu = (\mu_1, \ldots, \mu_{n-1})$ be two collections of integers.

$$\varphi(\lambda, \mu) = \int_{U(1)^\Gamma} h_{\ell_1}^{\lambda_1} \ldots h_{\ell_{2g}}^{\lambda_{2g}} h_{\partial F_1}^{\mu_1} \ldots h_{\partial F_{n-1}}^{\mu_{n-1}} d\nu_{x_1} \ldots d\nu_{x_p} dg'$$

$$= \int_{U(1)^\Gamma} h_{\lambda_1 \ell_1 + \ldots + \lambda_{2g} \ell_{2g} + \mu_1 \partial F_1 + \ldots + \mu_{n-1} \partial F_{n-1}} d\nu_{x_1} \ldots d\nu_{x_p} dg'$$

$$= \int_{U(1)^\Gamma} h_{e_1}^{\alpha_1} \ldots h_{e_r}^{\alpha_r} d\nu_{x_1} \ldots d\nu_{x_p} dg',$$

where the relation $\sum_i \lambda_i \ell_i + \sum_j \mu_j \partial F_j = \sum_k \alpha_k e_k$ defines $\alpha_1, \ldots, \alpha_r$. Suppose that the edges are labeled in such a way that $N_1 = e_1 \ldots e_{i_1}, \ldots, N_p = e_{i_{p-1}+1} \ldots e_{i_p}$ with $1 < i_1 < \ldots < i_p$. Then

$$\varphi(\lambda, \mu) = \int_{U(1)^\Gamma} (h_{e_1}^{\alpha_1} \ldots h_{e_{i_1}}^{\alpha_{i_1}}) \ldots (h_{e_{i_{p-1}+1}}^{\alpha_{i_{p-1}+1}} \ldots h_{e_{i_p}}^{\alpha_{i_p}}) h_{e_{i_p+1}}^{\alpha_{i_p+1}} \ldots h_{e_r}^{\alpha_r} d\nu_{x_1} \ldots d\nu_{x_p} dg'$$

$$= \int_{U(1)^{i_1}} g_1^{\alpha_1} \ldots g_{i_1}^{\alpha_{i_1}} d\nu_{x_1} \ldots \int_{U(1)^{i_p}} g_{i_{p-1}+1}^{\alpha_{i_{p-1}+1}} \ldots g_{i_p}^{\alpha_{i_p}} d\nu_{x_p}$$

$$\int_{U(1)} g^{\alpha_{i_p+1}} dg \ldots \int_{U(1)} g^{\alpha_r} dg.$$

This product is equal to zero as soon as there exists $k \geq i_p + 1$ such that $\alpha_k \neq 0$. We now claim that $\alpha_{i_p+1} = \ldots = \alpha_r = 0$ implies $\lambda = \mu = 0$.

Suppose that it is true. Then it implies that $\varphi(\lambda, \mu)$ is equal to 0 whenever $(\lambda, \mu) \neq (0, 0)$ and of course $\varphi(0, 0) = 1$. So $\varphi$ is equal to the characteristic function of the product of uniform measures on $U(1)^{2g+p-1}$ and the proposition is proved.

Let us prove our claim. If $M$ has no boundary, then $i_p + 1 = 1$, so that $\alpha_{i_p+1} = \ldots = \alpha_r = 0$ implies $\sum_i \lambda_i \ell_i + \sum_j \mu_j \partial F_j = 0$. But this equality read in $H_1(M; \mathbb{Z})$ gives $\lambda = 0$ and the fact that $\partial F_1, \ldots, \partial F_{n-1}$ are independent (see Proposition 1.33) then implies $\mu = 0$.

If $M$ has a boundary, then we can only say that the cycle $\sum \lambda_i \ell_i + \sum \mu_j \partial F_j$ is supported by $\partial M$. For some integers $\nu_1, \ldots, \nu_p$, we have the relation

$$\sum_{i=1}^{2g} \lambda_i \ell_i + \sum_{j=1}^{n-1} \mu_j \partial F_j = \sum_{k=1}^{p} \nu_k N_k,$$

which, in $H_1(M)$, gives $\nu_p[N_p] = \sum_i \lambda_i[\ell_i] - \sum_{k<p} \nu_k[N_k]$. Since $[N_p] = -\sum_{k<n}[N_k]$, this implies $\lambda = 0$ and $\nu_k = \nu_p$ for all $k$. Recall the relation $\sum_{j=1}^{n} \partial F_j = \sum_{i=1}^{p} N_i$, consequence of our choice of an orientation. We now have

$$\sum_{j=1}^{n-1} \mu_j \partial F_j = \nu_p \sum_{i=1}^{p} N_i = \nu_p \sum_{j=1}^{n} \partial F_j.$$

By Proposition 1.33 again, $\partial F_1, \ldots, \partial F_n$ are independent and the comparison of the coefficients of $\partial F_n$ gives $\nu_p = 0$, which implies $\mu = 0$ and hence terminates the proof. $\square$

There remains to study $h_{\partial F_n}$ in the fundamental system. With our convention on the orientation of $\partial F_1, \ldots, \partial F_n$ and $N_1, \ldots, N_p$, the relation

$$\partial F_n = \sum_{i=1}^{p} N_i - \sum_{j=1}^{n-1} \partial F_j$$

holds when $M$ has a boundary as well as when $M$ is closed. Thus, $h_{\partial F_n}$ is almost surely equal to $x_1 \ldots x_p h_{\partial F_1}^{-1} \ldots h_{\partial F_{n-1}}^{-1}$ under $\nu_{x_1} \otimes \ldots \otimes \nu_{x_p} \otimes dg'$, where, in the case $p = 0$, we adopt the usual convention that an empty product is equal to 1.

We can now summarize the results of this paragraph. Note that, in the Abelian setting, the measure $\nu_x^n$ is invariant under permutation of the factors in $U(1)^n$.

PROPOSITION 1.36. *Set $x = x_1 \ldots x_p$ if $M$ has a boundary and $x = 1$ if $M$ is closed. For any function $f$ continuous on $G^{2g+n+p}$, we have*

$$\int_{G^\Gamma} f(h_{\ell_1}, \ldots, h_{\ell_{2g}}, h_{N_1}, \ldots, h_{N_p}, h_{\partial F_1}, \ldots, h_{\partial F_n})\, dP_{(x_1, \ldots, x_p)} =$$

$$\frac{1}{p_{\sigma(M)}(x)} \int_{G^{2g+n}} f(u_1, \ldots, u_{2g}, x_1, \ldots, x_p, v_1, \ldots, v_n)\, p_{\sigma(F_1)}(v_1) \ldots p_{\sigma(F_n)}(v_n)$$

$$du_1 \ldots du_{2g}\, d\nu_x^n(v_1, \ldots, v_n).$$

The proof of the following corollary can be done using the same technique as in the proof of Proposition 1.35 and it is left to the reader.

COROLLARY 1.37. *If $c_1, \ldots, c_n$ are $n$ cycles such that the homology classes $[c_1], \ldots, [c_n]$ are linearly independent, then the random variables $h_{c_1}, \ldots, h_{c_n}$ are independent under $P_{(x_1, \ldots, x_p)}$.*

Together with Lemma 1.34, Proposition 1.36 certainly answers the question raised at the beginning of Section 1.9.1. But it is in fact possible to answer it using quite a different approach, which we are going to present now. It has the big advantage of being easily extended to the continuous situation.

### 1.9.3. Gaussian character of the Abelian theory.

The aim of this section is to give a new insight in the distribution of our fundamental system. Of course, there is not much more to say about the distribution of $(h_{\ell_1}, \ldots, h_{\ell_{2g}}, h_{N_1}, \ldots, h_{N_p})$ than what Proposition 1.36 already says. However, we are going to give a different description of the law of $(h_{\partial F_1}, \ldots, h_{\partial F_n})$ using the simple fact that the heat kernel on $U(1)$ is the image by the exponential mapping $t \mapsto e^{2i\pi t}$ of the heat kernel on $\mathbb{R}$. Let us begin with a short informal discussion.

Proposition 1.36 presents $h_{\partial F_1}, \ldots, h_{\partial F_n}$ as independent variables with respective distributions $p_{\sigma(F_1)}(x)\,dx, \ldots, p_{\sigma(F_n)}(x)\,dx$ conditioned by the fact that their product is equal to $x$. An alternative approach is to look for real-valued random variables which project via the exponential map on this distribution. If we forget the conditioning for one second, then a family of independent centered Gaussian variables with variances $\sigma(F_1), \ldots, \sigma(F_n)$ is certainly a good start. We have to condition them so that their sum projects on $x$. But there are many elements $t$ in $\mathbb{R}$ which project on a given $x \in U(1)$ by $t \mapsto e^{2i\pi t}$. A geometric argument may help us to understand this point. If we think of our real-valued variables as the random *curvature* of a random connection integrated over the faces $F_1, \ldots, F_n$ then their sum appears to play the role of the *total curvature*, and the total curvature of a connection on a principal $U(1)$-bundle depends only on the topology of the bundle when $M$ is closed. So we have to choose the topology of $P$, or at least, to choose a probability measure on the set of possible topological structures, which is in one-to-one correspondence with $\pi_1(U(1)) \simeq \mathbb{Z}$. This is the role of the variable $T$ in the following result, and a Gaussian-like distribution for this discrete parameter happens to be the good choice in order to recover the discrete measure. We will come back to this interpretation of $T$ in Section 3.2.

We keep our previous notation. In particular, $x$ is equal to the product $x_1 \ldots x_p$.

PROPOSITION 1.38. *Let $Y_1, \ldots, Y_n$ be independent real Gaussian random variables with $Y_i \sim \mathcal{N}(0, \sigma(F_i))$. Let $S = Y_1 + \ldots + Y_n$ be their sum. For each $i = 1, \ldots, n$, set*

$$X_i = Y_i - \frac{\sigma(F_i)}{\sigma(M)} S.$$

*Let $T$ be a real random variable, independent of the $Y_i$'s, with the following discrete distribution:*

$$\mathbb{P}(T = t) = \begin{cases} c \exp(-\frac{t^2}{2\sigma(M)}) & \text{if } e^{2i\pi t} = x \\ 0 & \text{otherwise,} \end{cases}$$

*where $c$ is an appropriate normalizing constant. Then, for any function $f$ continuous on $G^n$, we have*

$$\int_{G^\Gamma} f(h_{\partial F_1}, \ldots, h_{\partial F_n})\, dP_{(x_1, \ldots, x_p)} = \mathbb{E}\, f\left(e^{2i\pi\left(X_1 + \frac{\sigma(F_1)}{\sigma(M)}T\right)}, \ldots, e^{2i\pi\left(X_n + \frac{\sigma(F_n)}{\sigma(M)}T\right)}\right) \quad (1.20)$$

The proof of this proposition consists in an explicit verification.

PROOF. For this proof, we shall use the notation $\sigma_i = \sigma(F_i)$ and $\sigma_M = \sigma(M)$. Let us start by computing the right hand side of (1.20). One easily checks that

$$\mathbb{E} X_i X_j = \delta_{ij} \sigma_i - \frac{\sigma_i \sigma_j}{\sigma_M}$$

and $\sum X_i = 0$ a.s. The law of $(X_1, \ldots, X_n)$ has no density with respect to the Lebesgue measure on $\mathbb{R}^n$, but that of $(X_1, \ldots, X_{n-1})$ does, on $\mathbb{R}^{n-1}$. Denote by $C$ the $(n-1) \times (n-1)$ covariance matrix of $(X_1, \ldots, X_{n-1})$. It is easily checked that

$$(C^{-1})_{ij} = \frac{\delta_{ij}}{\sigma_i} + \frac{1}{\sigma_n},$$

so that the distribution of $(X_1, \ldots, X_{n-1})$ is equal to

$$\frac{1}{Z} \exp -\frac{1}{2} \left( \sum_{i=1}^{n-1} \frac{t_i^2}{\sigma_i} + \sum_{i,j=1}^{n-1} \frac{t_i t_j}{\sigma_n} \right) dt_1 \ldots dt_{n-1}.$$

Note that we can be loose about normalizing constants, because we know that we are dealing with probability measures. Let us fix a number $t_0$ such that $e^{2i\pi t_0} = x$. Let us also set $t_n = -t_1 - \ldots - t_{n-1}$. The right term of (1.20) is equal to

$$\frac{1}{Z} \sum_{k \in \mathbb{Z}} \int_{\mathbb{R}^{n-1}} f\left( e^{2i\pi(t_1 + \frac{\sigma_1}{\sigma_M}(t_0+k))}, \ldots, e^{2i\pi(t_{n-1} + \frac{\sigma_{n-1}}{\sigma_M}(t_0+k))}, e^{2i\pi(t_n + \frac{\sigma_n}{\sigma_M}(t_0+k))} \right)$$

$$\exp -\frac{1}{2} \left( \sum_{i=1}^{n-1} \frac{t_i^2}{\sigma_i} + \sum_{i,j=1}^{n-1} \frac{t_i t_j}{\sigma_n} \right) \exp -\frac{(t_0+k)^2}{2\sigma_M} dt_1 \ldots dt_{n-1}$$

$$= \frac{1}{Z} \sum_{k, q_1, \ldots, q_{n-1} \in \mathbb{Z}} \int_{[0,1]^{n-1}} f(e^{2i\pi t_1}, \ldots, e^{2i\pi t_{n-1}}, e^{2i\pi(t_n+t_0)})$$

$$\exp \left( -\frac{1}{2} \sum_{i=1}^{n-1} \frac{1}{\sigma_i} (t_i + q_i - \frac{\sigma_i}{\sigma_M}(t_0+k))^2 - \frac{1}{2\sigma_n} \left( \sum_{i=1}^{n-1} t_i + q_i - \frac{\sigma_i}{\sigma_M}(t_0+k) \right)^2 \right)$$

$$\exp -\frac{(t_0+k)^2}{2\sigma_M} dt_1 \ldots dt_{n-1}. \tag{1.21}$$

Now let us compute the left hand side of (1.20). For this, we need an expression of the heat kernel on $U(1)$. Recall that $U(1)$ is endowed with a metric of total volume equal to 1, so that the map $t \mapsto e^{2i\pi t}$ is a local isometry from $\mathbb{R}$ to $U(1)$, which sends the heat kernel on $\mathbb{R}$ to the heat kernel on $U(1)$. More explicitly, we have

$$p_s(e^{2i\pi t}) = \frac{1}{\sqrt{2\pi s}} \sum_{p \in \mathbb{Z}} \exp -\frac{(t-p)^2}{2s}. \tag{1.22}$$

We get

$$\int_{G^\Gamma} f(h_{\partial F_1}, \ldots, h_{\partial F_n}) \, dP_{(x_1, \ldots, x_n)} = \int_{G^n} f(v_1, \ldots, v_n) p_{\sigma_1}(v_1) \ldots p_{\sigma_n}(v_n) d\nu_x^n(v_1, \ldots, v_n)$$

$$= \frac{1}{Z} \sum_{p_1, \ldots, p_n \in \mathbb{Z}} \int_{[0,1]^{n-1}} f(e^{2i\pi t_1}, \ldots, e^{2i\pi t_{n-1}}, e^{2i\pi(t_n+t_0)}) \exp \left( -\frac{1}{2} \sum_{i=1}^{n-1} \frac{(t_i - p_i)^2}{\sigma_i} \right)$$

$$\exp \left( -\frac{1}{2} \frac{(t_n + t_0 - p_n)^2}{\sigma_n} \right) dt_1 \ldots dt_{n-1}. \tag{1.23}$$

By comparing (1.21) and (1.23), we see that the result will be a consequence of the following equality:

$$\sum_{p_1,\ldots,p_n \in \mathbb{Z}} \exp\left(-\frac{1}{2}\sum_{i=1}^{n-1}\frac{(t_i - p_i)^2}{\sigma_i} - \frac{1}{2}\frac{(t_n + t_0 - p_n)^2}{\sigma_n}\right) =$$

$$\sum_{k,q_1,\ldots,q_{n-1} \in \mathbb{Z}} \exp\frac{-(t_0+k)^2}{2\sigma_M} - \frac{1}{2}\sum_{i=1}^{n-1}\frac{1}{\sigma_i}(t_i + q_i - \frac{\sigma_i}{\sigma_M}(t_0+k))^2 -$$

$$-\frac{1}{2\sigma_n}\left(\sum_{i=1}^{n-1} t_i + q_i - \frac{\sigma_i}{\sigma_M}(t_0+k)\right)^2,$$

where $t_0, t_1, \ldots, t_{n-1} \in \mathbb{R}$ and $t_n = -t_1 - \ldots - t_{n-1}$. Setting $q_n = -q_1 - \ldots - q_{n-1}$, we have

$$\sum_{i=1}^{n-1}\frac{1}{\sigma_i}(t_i + q_i - \frac{\sigma_i}{\sigma_M}(t_0+k))^2 + \frac{1}{\sigma_n}(\sum_{i=1}^{n-1} t_i + q_i - \frac{\sigma_i}{\sigma_M}(t_0+k))^2 =$$

$$= \sum_{i=1}^{n-1}\frac{1}{\sigma_i}([t_i + q_i] - \frac{\sigma_i}{\sigma_M}(t_0+k))^2 +$$

$$+ \frac{1}{\sigma_n}\left([-t_n - q_n - (t_0+k)] + \frac{\sigma_n}{\sigma_M}(t_0+k)\right)^2$$

$$= \sum_{i=1}^{n-1}\frac{1}{\sigma_i}\left[(t_i + q_i)^2 - \frac{2\sigma_i}{\sigma_M}(t_i + q_i)(t_0+k) + \frac{\sigma_i^2}{\sigma_M^2}(t_0+k)^2\right] +$$

$$+ \frac{1}{\sigma_n}\left[(-t_n - q_n - (t_0+k))^2 + \frac{\sigma_n^2}{\sigma_M^2}(t_0+k)^2 + \right.$$

$$\left. + \frac{2\sigma_n}{\sigma_M}(-t_n - q_n - (t_0+k))(t_0+k)\right]$$

$$= \sum_{i=1}^{n-1}\frac{1}{\sigma_i}(t_i + q_i)^2 + \frac{1}{\sigma_M}(t_0+k)^2\left(\frac{\sigma_n}{\sigma_M} + \sum_{i=1}^{n-1}\frac{\sigma_i}{\sigma_M}\right)$$

$$- \frac{2}{\sigma_M}\sum_{i=1}^{n-1}(t_i + q_i)(t_0+k) + \frac{2}{\sigma_M}\sum_{i=1}^{n-1}(t_i + q_i)(t_0+k)$$

$$- \frac{2}{\sigma_M}(t_0+k)^2 + \frac{1}{\sigma_n}(-t_n - q_n - (t_0+k))^2$$

$$= \sum_{i=1}^{n-1}\frac{1}{\sigma_i}(t_i + q_i)^2 + \frac{1}{\sigma_n}(t_n + t_0 + q_n + k)^2 - \frac{1}{\sigma_M}(t_0+k)^2.$$

Setting $p_1 = -q_1, \ldots, p_{n-1} = -q_{n-1}$ and $p_n = -q_n - k$, we get the result. □

**1.9.4. The double layer potential.** For the moment, Proposition 1.38 looks more like a computational trick than a real progress in the understanding of the discrete Yang-Mills measure. But we claim that the Gaussian variables $(X_1, \ldots, X_n)$ are in fact naturally attached to the loops $\partial F_1, \ldots, \partial F_n$, and in a way which can be generalized to a very wide class of loops. More precisely, we claim that $(X_1, \ldots, X_n)$

is isometric as a vector of elements of $L^2(\mathbb{P})$ to a vector of functions of $L^2(M, \sigma)$ which is itself naturally attached to the family of loops $(\partial F_1, \ldots, \partial F_n)$.

For this section, we choose an orientation on $M$ and we assume that all boundaries are oriented accordingly.

To begin with, observe that the vector $(\mathbf{1}_{F_1}, \ldots, \mathbf{1}_{F_n}, 1)$ of $L^2(M, \sigma)$ is isometric to $(Y_1, \ldots, Y_n, S)$ defined in Proposition 1.38, because $\mathbb{E}(Y_i Y_j) = (\mathbf{1}_{F_i}, \mathbf{1}_{F_j})_{L^2} = \delta_{i,j} \sigma(F_i)$. Now if we set

$$u_i = \mathbf{1}_{F_i} - \frac{\sigma(F_i)}{\sigma(M)},$$

then the vector $(u_1, \ldots, u_n)$ is clearly isometric to $(X_1, \ldots, X_n)$. For each $i$, $\mathbf{1}_{F_i}$ is a naive version of the winding number of the loop $\partial F_i$ and $u_i$ is the orthogonal projection of $\mathbf{1}_{F_i}$ on the hyperplane $L_0^2(M, \sigma)$ of zero-mean functions.

Let us give a direct definition of each $u_i$ as the double-layer potential of the loop $\partial F_i$. This is the correct generalization of the winding number for a loop on a Riemannian surface.

To do this, we need indeed to endow $M$ with a Riemannian metric which is consistent with the measure $\sigma$ in that its Riemannian volume measure is equal to $\sigma$. Such a metric exists because $\sigma$ is Lebesguian (see 1.1), and in fact any metric on $M$ is conformally equivalent to a metric with Riemannian volume $\sigma$.

The choice of a compatible metric on $M$ gives rise to a Laplace operator $\Delta$ on functions and, since an orientation has been fixed, to a Hodge operator $*$ on the space $\Omega^1(M)$ of differential 1-forms. The Laplace operator has a Green function (see [8, 42]) $G : M \times M \longrightarrow \mathbb{R}_+$, which is a smooth symmetric function defined outside the diagonal and such that for any function $\phi \in C^2(M)$ and for any $x \in M$,

$$\phi(x) - \frac{1}{\sigma(M)} \int_M \phi(y) \, d\sigma(y) = \int_M G(x, y) \Delta \phi(y) \, d\sigma(y). \tag{1.24}$$

Moreover, it satisfies

$$\int_M G_x \, d\sigma = 0 \quad \forall x \in M,$$

where $G_x$ denotes the function on $M$ defined by $G_x(y) = G(x, y)$. When $M$ has a boundary, we choose to consider the Green function with Neumann boundary conditions, which we also denote by $G$. In this case, Equation (1.24) must be replaced by

$$\phi(x) - \frac{1}{\sigma(M)} \int_M \phi(y) \, d\sigma(y) = \int_M G(x, y) \Delta \phi(y) \, d\sigma(y) - \int_{\partial M} G_x * d\phi, \tag{1.25}$$

and one has the boundary condition

$$j^*(*dG_x) = 0 \quad \forall x \in M,$$

where $j : \partial M \longrightarrow M$ denotes the canonical embedding.

Finally, the divergence near the diagonal of the Green function is logarithmic. More precisely, the function

$$(x, y) \longmapsto G(x, y) - \frac{1}{2\pi} \log d(x, y)$$

is smooth on $M \times M$.

DEFINITION 1.39. Let $c$ be a path on $M$. We call *double layer potential* of $c$ the function $u_c$ defined on $M$ outside the image of $c$ by :
$$u_c(x) = \int_c *dG_x.$$

Note that the double layer potential of any element $c = \sum_k \alpha_k e_k$ of $\mathbb{Z}^\Gamma$ is well defined by $u_c = \sum_k \alpha_k u_{e_k}$ and that this extension is additive in the sense that if $c_1$ and $c_2$ are two elements of $\mathbb{Z}^\Gamma$, then $u_{c_1+c_2} = u_{c_1} + u_{c_2}$ $\sigma$-a.e. on $M$.

PROPOSITION 1.40. *Let $c$ be a simple loop which is the oriented boundary of an open subset $U$ of $M$. Set $V = U^c$. Then*
$$u_c = \frac{\sigma(V)}{\sigma(M)}\mathbf{1}_U - \frac{\sigma(U)}{\sigma(M)}\mathbf{1}_V = \mathbf{1}_U - \frac{\sigma(U)}{\sigma(M)}.$$

*In particular, $u_c \in L^2(M, \sigma)$ and $\| u_c \|_2 = \left(\frac{\sigma(U)\sigma(V)}{\sigma(M)}\right)^{\frac{1}{2}}$.*

PROOF. Let $x$ be in $U$. Since $*dG_x = 0$ along $\partial M$ whenever $M$ has a boundary and $\Delta G_x = -\frac{1}{\sigma(M)}$ on $M - \{x\}$, we have:
$$u_c(x) = \int_{\partial U} *dG_x = -\int_{\partial V} *dG_x = -\int_V -\frac{1}{\sigma(M)} d\sigma = \frac{\sigma(V)}{\sigma(M)}.$$
Now let $x$ be in $V$.
$$u_c(x) = \int_{\partial U} *dG_x = \int_U -\frac{1}{\sigma(M)} d\sigma = -\frac{\sigma(U)}{\sigma(V)}.$$
The last part of the statement follows easily. □

In what follows, we identify functions on $M$ which are equal almost everywhere.

COROLLARY 1.41. *The vector $(u_1, \ldots, u_n)$ is equal to $(u_{\partial F_1}, \ldots, u_{\partial F_n})$.*

In order to go from the functions $u_{\partial F_i}$ to the random variables $X_i$, we need an isometry from $L^2(M, \sigma)$ into a Gaussian subspace of $L^2(\mathbb{P})$. Such an isometry is called a *white noise* on $(M, \sigma)$. Let us fix a linear mapping
$$\begin{aligned} W : L^2(M, \sigma) &\longrightarrow L^2(\mathbb{P}) \\ u &\longmapsto W(u) \end{aligned}$$
such that for any $u, v \in L^2(M)$, $W(u)$ and $W(v)$ are centered real Gaussian random variables with covariance $\mathbb{E}[W(u)W(v)] = (u, v)_{L^2}$. Proposition 1.38 can be rewritten in the following form:

PROPOSITION 1.42. *The following equality holds in distribution:*
$$(h_{\partial F_1}, \ldots, h_{\partial F_n}) \stackrel{(d)}{=} \left(e^{2i\pi\left(W(u_{\partial F_1}) + \frac{\sigma(F_1)}{\sigma(M)}T\right)}, \ldots, e^{2i\pi\left(W(u_{\partial F_n}) + \frac{\sigma(F_n)}{\sigma(M)}T\right)}\right).$$

This achieves the first goal of this section, namely to reformulate Proposition 1.38 in a more natural way. It is now fairly easy to generalize Proposition 1.42 to an arbitrary family of homologically trivial cycles.

Let $c_1, \ldots, c_k \in C_0\Gamma$ be such cycles. For each $i$, we know by Proposition 1.33 that $c_i$ is a linear combination of the loops $\partial F_i$, so that, by the last statement of Proposition 1.40, $u_{c_i}$ belongs to $L^2(M)$, and finally $W(u_{c_i})$ is well-defined.

It remains only to figure out the correct generalization of the term $\frac{\sigma(F_i)}{\sigma(M)}$. Our choice of an orientation of $M$ allows us to identify $\sigma$ with a 2-form on $M$. If $M$ has a boundary, then $H^2(M) = 0$ so that there exists a 1-form $\gamma$ such that $d\gamma = \sigma$. This form is defined up to addition of a closed 1-form and each $c_i$ is homologically trivial, so that the numbers

$$\sigma(c_i) := \frac{1}{\sigma(M)} \int_{c_i} \gamma \qquad (1.26)$$

are well-defined.

If $M$ is closed, let us consider $c_1$ for example. There is no 1-form whose differential equals $\sigma$, but if $m$ is any point of $M$ chosen outside the range of $c_1$, then $H^2(M - \{m\}) = 0$ and there exists $\gamma \in \Omega^1(M - \{m\})$ such that $d\gamma = \sigma$. We claim that the class modulo 1 of the number $\sigma(c_1)$ defined as in (1.26) is independent of the choice of $m$. If $m'$ is another point outside the range of $c_1$ and $\gamma'$ a corresponding 1-form, we want to show that

$$\int_{c_1} \gamma - \gamma' \in \mathbb{Z}.\sigma(M). \qquad (1.27)$$

Since $\gamma - \gamma'$ is a closed form on $M - \{m, m'\}$, this integral depends on $c_1$ only through its homology class in $H_1(M - \{m, m'\})$. Let $l$ be the boundary of a small disk $D \subset M$ around $m'$ which does not contain $m$. Then $H_1(M - \{m, m'\})$ is generated by the classes of a system of representatives of a basis of $H_1(M)$ and the class of $l$. Since $[c] = 0$ in $H_1(M)$, $[c] = p[l]$ for some $p \in \mathbb{Z}$ in $H_1(M - \{m, m'\})$. There remains to compute $\int_l \gamma - \gamma'$:

$$\int_l \gamma - \gamma' = \int_D \sigma + \int_{M-D} \sigma = \sigma(M).$$

This establishes (1.27). In this case, the number $\sigma(c_i)$ is only defined modulo 1, but at the same time, $T$ takes its values in $\mathbb{Z}$ because $x = 1$, so that $\exp(2i\pi\sigma(c_i)T)$ is well-defined anyway.

PROPOSITION 1.43. *Let $(c_1, \ldots, c_k)$ be cycles of $C_0\Gamma$. Then the following equality holds in distribution:*

$$(h_{c_1}, \ldots, h_{c_k}) \stackrel{(d)}{=} \left(e^{2i\pi\left(W(u_{c_1}) + \sigma(c_1)T\right)}, \ldots, e^{2i\pi\left(W(u_{c_k}) + \sigma(c_k)T\right)}\right).$$

PROOF. We already know by Proposition 1.42 that the result is true when $(c_1, \ldots, c_k) = (\partial F_1, \ldots, \partial F_n)$. Now, by Proposition 1.33, the boundaries of the faces generate the $\mathbb{Z}$-module $C_0\Gamma$, so that it is sufficient to show that the new set of variables defined using the white noise enjoys the same multiplicativity property as $(h_c)_{c \in C_0\Gamma}$.

On one hand, $W$ is linear and the double layer potential is additive, so that $\exp 2i\pi W(u_{c_1+c_2}) = \exp 2i\pi W(u_{c_1}) \exp 2i\pi W(u_{c_2})$.

On the other hand, $\sigma(c_1 + c_2) = \sigma(c_1) + \sigma(c_2)$, at least modulo 1 when $M$ is closed, and the result is proved. □

As explained at the beginning of Section 1.9, this investigation will be completed in Chapter 3 in the light of the construction of the continuous Yang-Mills measure, which is the object of the next chapter.

Some of the techniques used for this construction will allow us to establish important properties of the double layer potential but we need to prove here the fundamental one.

THEOREM 1.44. *If $M$ is closed, then the double layer potential of any path $c$ of $PM$ belongs to $L^\infty(M,\sigma)$. In particular it belongs to $L^2(M,\sigma)$.*

Let us begin by what looks like a very special case.

PROPOSITION 1.45. *There exists $L > 0$ such that for all geodesic path $\zeta \in PM$ such that $\ell(\zeta) < L$, the double layer potential $u_\zeta$ of $\zeta$ belongs to $L^\infty(M)$.*

The idea of the proof is to show that the form $*dG_x$, up to a smooth remainder, can be compared with the natural angular form $d\theta_x$ in the neighbourhood of each $x$. Then, one concludes by observing that the angle under which one sees a short geodesic from a point close to it is always less than $\pi$.

PROOF. The proof relies on the fact that we know quite precisely the behaviour of the Green function near the diagonal. In fact, if $R$ denotes the radius of injectivity and if $f$ is a smooth function on $[0,\infty)$ such that $f(r) = 1$ if $r \leq R/2$ and $f(r) = 0$ if $r \geq R$, then the Green function can be written

$$G(x,y) = -\frac{1}{2\pi} f(d(x,y)) \log d(x,y) + F(x,y)$$

where $F$ is smooth on $M \times M$ (see the construction of $G$ in [8]). The part of the double layer potential of a path corresponding to the integral of $*dF$ is smooth on $M$, so that we are really concerned only by the logarithmic part.

Let $x$ be fixed for one moment and let us consider the differential form $*d\log r$ in the geodesic ball $B(x, R/2)$, where $r(y) = d(x,y)$. In $B(x, R/2) - \{x\}$, let us define the two vector fields $\partial_r$ and $\partial_\theta$ by

$$\partial_r(\exp v) = \frac{1}{\|v\|} \left.\frac{d}{dt}\right|_{t=1} \exp(tv) \quad \text{and} \quad \partial_\theta(\exp v) = \left.\frac{d}{d\theta}\right|_{\theta=0} \exp(R_\theta(v)),$$

where $v \in T_x M$, $\|v\| < R/2$, and $R_\theta$ denotes the rotation of angle $\theta$ in $T_x M$. Observe that $\|\partial_r\| = 1$ and that $\partial_r$ and $\partial_\theta$ are orthogonal. We can define the forms $dr = \langle \partial_r, \cdot \rangle$ and $d\theta = \|\partial_\theta\|^{-2} \langle \partial_\theta, \cdot \rangle$, so that $(dr, d\theta)$ is the dual basis of $(\partial_r, \partial_\theta)$. With this notation, we have

$$*d\log r = \frac{*dr}{r} = \frac{\|\partial_\theta\|}{r} d\theta.$$

We need the following lemma.

LEMMA 1.46. *Let $C$ denote the function defined in a neighbourhood of the zero section in $TM$ by*

$$C(x,v) = \frac{\|\partial_{\theta_x}(v)\|}{\|v\|}.$$

*There exists $R' < R/2$ and $K$ such that $0 \leq C(x,v) \leq K$ whenever $\|v\| \leq R'$.*

PROOF. Let us consider the differential of the exponential map $d\exp : TTM \longrightarrow TM$, actually its restriction to the subbundle $T^V TM$ of $TTM$ consisting of vertical vectors. So we are considering the mapping

$$d\exp : T^V TM \longrightarrow TM,$$

where $d_v \exp : T_v^V T_x M \longrightarrow T_{\exp_x v} M$ if $v \in T_x M$. For each $v \in TM$, the canonical identification $T_v^V T_x M \simeq T_x M$ allows us to endow each fiber of $T^V TM$ with a scalar product. In this way, $d_v \exp$ becomes a linear map between two Euclidean spaces whose Hilbert-Schmidt norm $\| d_v \exp \|_2$ is well-defined. Moreover, this norm depends continuously on $v$, for we can choose in the neighbourhood $U$ of any point $x$ of $M$ a system of two orthonormal vector fields $X_1$ and $X_2$, so that, if $v$ is close enough to 0 that $\exp_x v \in U$,

$$\| d_v \exp \|_2^2 = \sum_{i,j=1,2} <d_v \exp X_i, X_j>^2$$

which is continuous. Finally, according to the definition of $\partial_\theta$, we have

$$C(x,v) \leq \sup_{w \in T_v^V T_x M} \frac{\| d_v \exp(w) \|}{\| w \|} \leq \| d_v \exp \|_2,$$

so that $C(x,v)$ is bounded by a continuous function for $v$ small enough. This proves that there exists $R'$, that we can choose smaller than $R/2$, and $K$ such that $0 \leq C(x,v) \leq K$ whenever $\| v \| \leq R'$. $\square$

Consider now a geodesic path $\zeta$ such that $\ell(\zeta) < R'/4$ and let $B$ be the open ball of radius $R'/2$ around $\zeta(0)$. Any point outside $B$ is at a distance greater than $R'/4$ of $\zeta$. Since the Green function is smooth on $\{(x,y) \in M \times M : d(x,y) > R'/4\}$, the double layer potential of $\zeta$ is smooth, hence bounded on $M - B$. Now consider $x$ in $B$. The path $\zeta$ is contained in the ball $B(x, R')$, so that we can work in normal coordinates at $x$ and use the lemma above. Let us fix a parametrization of $\zeta$ and define $v : [0,1] \longrightarrow T_x M$ such that $\exp_x v(t) = \zeta(t)$. We find

$$\int_\zeta *d \log r_x = \int_0^1 C(x, v(t)) \, d\theta_x(\dot\zeta(t)).$$

In order to be able to use our lemma, we need a positivity argument. And this is where the fact that $\zeta$ is geodesic is essential. Indeed, we claim that the sign of $d\theta_x(\dot\zeta(t))$ is constant. If there exists $t_0$ such that $d\theta_x(\dot\zeta(t_0)) = 0$, then $\zeta$ is tangent to the geodesic through $x$ and $\zeta(t_0)$. This implies that $\zeta$ is a portion of this geodesic and hence that $d\theta_x(\dot\zeta(t))$ is identically equal to zero. Hence, we can deduce the relation

$$\left| \int_\zeta *d \log r \right| \leq K \left| \int_\zeta d\theta_x \right|.$$

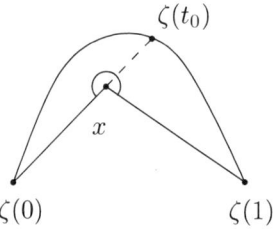

FIGURE 3. If the small geodesic triangle $(x, \zeta(0), \zeta(1))$ is not flat and has an angle greater than $\pi$ at $x$, then one can find $t_0$ such that $\zeta$ is not minimizing on $[0, t_0]$.

Finally, we need to bound the right hand side of this inequality. However, just as in the euclidean case, it is at most equal to $\pi$, because every angle of a small geodesic triangle is less than $\pi$ (see Figure 3).

This shows that $|u_\varsigma|$ is bounded by $K\pi$ inside $B = B(\varsigma(0), R'/2)$ and terminates the proof. $\square$

PROOF OF THEOREM 1.44. First of all, Proposition 1.45 extends immediately to geodesic segments of arbitrary length. Then, we need to use a result proved in Chapter 2, namely Proposition 2.29, which shows in particular that any embedded path can be completed into a simple homotopically trivial loop by concatenating it with a finite number of geodesic segments. Now the double layer potential of such a loop is bounded according to Proposition 1.40.

This shows that the double layer potential of any embedded path is bounded. The general case follows by additivity. $\square$

CHAPTER 2

# Continuous Yang-Mills measure

This chapter is the heart of our study. It will provide us with a pathwise multiplicative random process indexed by the set of loops on $M$ which we shall regard as the parallel transport for a random connection distributed according to the Yang-Mills measure.

The first step of the construction is an extension of the discrete construction which allows us, given a Riemannian metric on $M$, to define simultaneously a random variable for each piecewise geodesic path. This is the content of the first two sections. The next step is to prove an approximation result and to extend the random parallel transport to the set of paths on $M$ defined in Section 1.2.1. Finally, one needs to check that this extension is consistent with the discrete theory. Sections 2.4 to 2.6.4 are devoted to the approximation result and to the extension of the set of random variables. Consistency is checked in Section 2.7, then Section 2.9 deals with the case of surfaces with boundary and the rest of the chapter puts these results together in order to define the Yang-Mills measure properly. Theorem 2.58 is the result around which the whole chapter is organized and it could be helpful to read it now.

Two properties of the discrete theory are really essential to our purpose, namely the subdivision invariance, as expressed in Theorem 1.22, and the regularity property stated in Proposition 1.29.

## 2.1. Projective systems of probability spaces

Put in its simplest form, our basic idea is to cover $M$ with an increasingly fine sequence of graphs and to prove that the sequence of discrete measures associated with these graphs converges in some sense to a continuous object. Although the subdivision invariance property strongly supports the belief that such a procedure is workable, it appears that it cannot be made rigorous in this primitive form. To see why, and also because this will be useful in the sequel, let us describe the natural framework in which our first idea could be given a precise meaning, namely that of projective limits of probability spaces.

Let us begin by defining a projective system of probability spaces.

DEFINITION 2.1. Let $(\Lambda, \leq)$ be a directed set, that is, a partially ordered set such that for all $\lambda, \mu \in \Lambda$, there exists $\nu \in \Lambda$ such that $\lambda \leq \nu$ and $\mu \leq \nu$. A *projective family of probability spaces indexed by* $\Lambda$ is a family $(\Omega_\lambda, \mathcal{B}_\lambda, P_\lambda)$ of probability spaces together with a family of measurable maps $f_{\lambda\mu} : \Omega_\mu \longrightarrow \Omega_\lambda$ defined for all $\lambda \leq \mu$, such that for all $\lambda \leq \mu \leq \nu \in \Lambda$,
 1. $f_{\lambda\lambda} = \mathrm{Id}_{\Omega_\lambda}$,
 2. $f_{\lambda\nu} = f_{\lambda\mu} \circ f_{\mu\nu}$,
 3. $(f_{\lambda\mu})_* P_\mu = P_\lambda$.

Note that we do not assume the surjectivity of the maps $f_{\lambda\mu}$. The two examples of directed sets that we have in mind are a class of graphs on $M$ ordered by the relation of finess and the set of finite subsets of $PM$ ordered by inclusion.

The set-theoritic *projective limit* of a projective system is by definition the set

$$\varprojlim \Omega_\lambda = \Omega = \{(\omega_\lambda)_{\lambda \in \Lambda} \in \prod_{\lambda \in \Lambda} \Omega_\lambda : f_{\lambda\mu}(\omega_\mu) = \omega_\lambda \ \forall \lambda \leq \mu\}.$$

For each $\lambda \in \Lambda$, the projection on the $\lambda$-th coordinate defines a map $f_\lambda : \Omega \longrightarrow \Omega_\lambda$. These maps allow one to define the algebra of subsets $\mathcal{B}_0 = \bigcup_\lambda f_\lambda^{-1}(\mathcal{B}_\lambda)$ on $\Omega$ and the compatibility conditions of the projective system ensure that the relation $P_0(A) = P_\lambda(B)$ if $A = f_\lambda^{-1}(B)$ defines without ambiguity an additive measure on $\mathcal{B}_0$. It is not true in general that $P_0$ can be extended to a $\sigma$-additive measure $P$ on $\mathcal{B} = \sigma(\mathcal{B}_0)$, nor even, when this extension exists, that $(f_\lambda)_* P = P_\lambda$ for every $\lambda^1$. What will protect us from this pathology is the compacity of our probability spaces, via the following theorem. The notation is that introduced above.

THEOREM 2.2. *Assume that each measurable space $(\Omega_\lambda, \mathcal{B}_\lambda)$ is a compact Hausdorff topological space endowed with its Borel $\sigma$-algebra, and that each map $f_{\lambda\mu}$ is continuous. Then, for each $\lambda \in \Lambda$,*

$$f_\lambda(\Omega) = \bigcap_{\lambda \leq \mu} f_{\lambda\mu}(\Omega_\mu).$$

*In particular, $f_\lambda(\Omega)$ contains the support of $P_\lambda$.*

*Moreover, the additive measure $P_0$ on $\mathcal{B}_0$ can be extended to a $\sigma$-additive measure $P$ on $\mathcal{B}$ and this measure satisfies $(f_\lambda)_* P = P_\lambda$ for each $\lambda$. Finally, the measurable space $(\Omega, \mathcal{B})$ is itself a compact Hausdorff space endowed with its Borel $\sigma$-algebra.*

The first half of this theorem is proved in [15] and the second in [35]. It would be tempting to take for $\Lambda$ the set of all graphs on $M$ and to claim that the projective limit provided by Theorem 2.2 supports a family of random variables indexed by the set of all paths on $M$. The problem is that given two graphs, it is not the case in general that there exists a third one which is finer than both others, because two edges belonging to two different graphs can intersect very badly, even though each of them is very regular: the images of two edges can for example determine an infinity of faces on $M$ and this is out of the reach of the discrete theory.

Thus, we must consider a smaller family of graphs, one in which any two elements are always dominated by a third one. A. Ashtekar and J. Lewandowski proposed in [6] to use graphs with piecwise real analytic edges, for some complex structure on $M$. We are rather going to use piecewise geodesic edges for some Riemannian metric on $M$, but the whole procedure of approximation, could presumably be done in the same way with analytic edges. The point is to take edges in such a class of 'rigid' curves that two edges can cross each other only a finite number of times.

Another strategy consists in analyzing the possible pathology of the intersection of two edges. This has been done by J. Baez and S. Sawin, using *tassels* and *webs*, in [9, 10]. They have been able to use this analysis to construct the kinematical measure, that is, the formal measure $dA$ with respect to which the Yang-Mills

---

[1]In general, it can happen that $\Omega$ is *empty*, even when all the maps $f_{\lambda\mu}$ are onto !

measure is informally supposed to have the smooth density $\exp -S(A)/2^2$. As far as I know, they have not generalized their approach to the construction of the Yang-Mills measure itself.

## 2.2. A Riemannian metric

Recall that we are considering a compact surface $M$ endowed with a Lebesguian measure $\sigma$, one which is equivalent to the Lebesgue measure in any chart. For technical reasons we will assume, unless explicitly stated, that $M$ has no boundary until Section 2.9, where we shall derive the construction in the general case from that in the case of closed surfaces. The reason for this restriction is that we are going to use tubular neighbourhoods of embedded paths, which do not always exist on manifolds with boundary.

Let us choose $q$ disjoint simple loops $L_1, \ldots, L_q$ on $M$ whose ranges are smooth submanifolds of $M$, in other words, $q$ disjoint embeddings of $S^1$ into $M$. Note that $(L_1, \ldots, L_q)$ is in particular a disintegration family in the sense of Remark 1.17. We will sometimes think of these loops as the boundary of a submanifold of $M$ (and this points to the construction of the measure on surfaces with boundary) or just as loops along which we want to put contitions on the holonomy.

As explained at the end of the previous section, we are going to consider piecewise geodesic loops for some Riemannian metric on $M$. Just as in Section 1.9.4, we want this metric to be compatible with the only geometric structure we have on $M$, namely the surface measure $\sigma$. But we want also to be able to take the paths $L_1, \ldots, L_q$ into account in any discrete-theoritic computation involving piecewise geodesic paths. For this reason, we want $L_1, \ldots, L_q$ to be geodesic loops for our metric. This section is devoted to the proof of the following existence result.

PROPOSITION 2.3. *There exists a Riemannian metric on $M$ whose Riemannian volume coincides with $\sigma$ and such that $L_1, \ldots, L_q$ are geodesics.*

We are going to adapt a proof of the following classical theorem.

THEOREM 2.4 (Moser [33]). *Let $\alpha$ and $\beta$ be two volume 2-forms on a closed oriented compact surface $M$ such that*

$$\int_M \alpha = \int_M \beta.$$

*There exists a diffeomorphism $\phi : M \longrightarrow M$ such that $\phi^* \beta = \alpha$.*

PROOF OF PROPOSITION 2.3. If $q=0$, let us choose an arbitrary metric on $M$. Its Riemannian volume is equivalent to $\sigma$, with a smooth density. Multiplying the metric by an appropriate smooth positive function, we can produce a new metric, conformally equivalent to the first one, whose Riemannian volume is equal to $\sigma$.

If $q > 0$, the proof is more complicated. Let us first construct a metric for which $L_1, \ldots, L_q$ are geodesic. Each $L_i$ has a tubular neighbourhood in $M$ which is diffeomorphic to a cylinder $S^1 \times (-1,1) \ni (\theta, t)$, and in which $L_i$ is defined by

---

[2]It is however easy to check that the kinematical and Yang-Mills measures are mutually singular, by considering for example the random variables associated to a sequence of small circles centered at a given point. Under the measure constructed by Ahstekar and Lewandowski, all these variables are uniformly distributed on $G$ and independent. On the other hand, if the radii of the circles tend fast enough to 0, then the same random variables tend to 1 almost surely for the Yang-Mills measure. See also [24].

the equation $t = 0$. In these coordinates, $L_i$ is certainly geodesic for the metric $d\theta^2 + dt^2$. If we choose the tubular neighbourhoods small enough, these metrics are defined on subsets which do not overlap, and we can extend them to a metric $g_0$ on $M$.

Observe that the loops $L_1, \ldots, L_q$ split $M$ into several submanifolds with boundary. More properly speaking, there exist $k$ manifolds with boundary and $k$ immersions $j_i : M_i \longrightarrow M$, $i = 1 \ldots k$, such that each $j_i$ is an embedding of the interior of $M_i$ and $M - N = \cup_i j_i(\overset{\circ}{M_i})$, with $N = \cup_i L_i(S^1)$.

By multiplying $g_0$ by an appropriate positive function identically equal to 1 in a neighbourhood of each $L_i$, we can produce a new metric $g_1$ for which each $L_i$ is still geodesic and also such that

$$\text{vol}_{g_1}(M_i) = \sigma(M_i) \quad \forall i = 1 \ldots k. \tag{2.1}$$

Now we must redistribute the total area inside each submanifold $M_i$. To do this, we adapt the proof of Moser's theorem 2.4.

Let us choose two 2-forms $\alpha$ and $\beta$ on $M$ representing respectively $\sigma$ and $\text{vol}_{g_1}$, after a choice of orientation of $M$. We know more about these 2-forms than what is needed to apply Moser's theorem, but we also want to prove more: we want to find a diffeomorphism of $M$ which sends $\alpha$ to $\beta$ and also preserves the loops $L_1, \ldots, L_q$, so that they are still geodesic for the pull-back of $g_1$ by this diffeomorphism.

Consider $N = \cup_i L_i(S^1)$ and $j : N \longrightarrow M$ the canonical injection. Since $M$ is closed, $H^2(M; \mathbb{R}) \simeq \mathbb{R}$ and the isomorphism is given by $\eta \mapsto \int_M \eta$. So the fact that $\int_M \beta - \alpha = 0$ implies that there exists a form $\gamma \in \Omega^1(T^*M)$ such that $d\gamma = \beta - \alpha$. We claim that $\gamma$ can be chosen such that $j^*\gamma = 0$, or in other words such that $\gamma(X) = 0$ for any vector $X$ tangent to a loop $L_i$. Equivalently, we claim that the class of $\beta - \alpha$ in the relative cohomology group $H^2(M, N)$ is zero.

To see this, we prove that this class is equal to zero as an element of the dual of $H_2(M, N)$. For each $i = 1, \ldots, k$, the fundamental class of $M_i$ exists in $H_2(M_i, \partial M_i)$ and its image by $j_{i*}$ defines a class in $H_2(M, N)$, on which $[\beta - \alpha]$ vanishes by (2.1). So it suffices to prove that $H_2(M, N)$ is generated by the $j_{i*}[M_i]$.

Let $X$ denote the disjoint union of $M_1, \ldots, M_k$ and $A$ the subset of $X$ formed by the union of their boundaries. Then $A$ is a strong deformation retract of one of its closed neighbourhoods in $X$, for example the reunion of one closed half-tube around each boundary component (see Section 2.5.2 for a more detailed argument). Finally, the map $f$ from $X$ to $M$ induced by $j_1, \ldots, j_k$ sends $A$ on $N$ and maps $X - A$ homeomorphically onto $M - N$. Then Theorem 4.8.9 of [41] asserts that $f$ induces an isomorphism between the relative homology groups of $(X, A)$ and $(M, N)$. In particular, since $H_2(X, A) = \bigoplus H_2(M_i, \partial M_i)$ is generated by the fundamental classes of the $M_i$'s, $H_2(M, N)$ is generated by the classes $j_{i*}[M_i]$.

We have proved that it is possible to choose $\gamma$ such that $\gamma(X) = 0$ for each vector $X$ tangent to an $L_i$, and we do choose it in that way. The end of the proof is now similar to that of Moser's theorem.

For each $t \in [0, 1]$, set $\alpha_t = (1 - t)\alpha + t\beta$ and define the vector field $X_t$ on $M$ by $\iota_{X_t}\alpha_t = -\gamma$. This makes sense because, since $\alpha$ and $\beta$ determine the same orientation of $M$, $\alpha_t$ is a volume forme for each $t$. The field $X_t$ depends smoothly on $t$ and induces a flow $(\varphi_t)_{t \in [0,1]}$, that is, a semi-group of diffeomorphisms of $M$

such that
$$\left.\frac{d}{dt}\right|_{t=t_0} \varphi_t(m) = X_{t_0}(m) \quad \forall t_0 \in [0,1], \ \forall m \in M.$$

Let us compute the derivative of $\varphi_t^* \alpha_t$. For any $t_0 \in [0,1]$,
$$\left.\frac{d}{dt}\right|_{t=t_0} (\varphi_t^* \alpha_t) = \left.\frac{d}{dt}\right|_{t=t_0} (\varphi_t^* \alpha_{t_0}) + \left.\frac{d}{dt}\right|_{t=t_0} (\varphi_{t_0}^* \alpha_t).$$

The second term of the r.h.s. is equal to
$$\varphi_{t_0}^* \left( \left.\frac{d}{dt}\right|_{t=t_0} \alpha_t \right) = \varphi_{t_0}^*(\beta - \alpha).$$

Let us denote by $\mathcal{L}_{X_t}$ the Lie derivative with respect to the field $X_t$ and use Cartan's relation $\mathcal{L}_{X_t} = d \circ \iota_{X_t} + \iota_{X_t} \circ d$. The first term of the r.h.s. is equal to
$$\mathcal{L}_{X_{t_0}} \varphi_{t_0}^* \alpha_{t_0} = d\left(\iota_{X_{t_0}} \varphi_{t_0}^* \alpha_{t_0}\right) = d(\varphi_{t_0}^* \iota_{X_{t_0}} \alpha_{t_0}) = -d(\varphi_{t_0}^* \gamma) = -\varphi_{t_0}^*(d\gamma) = -\varphi_{t_0}^*(\beta - \alpha).$$

Thus, $\frac{d}{dt}\varphi_t^* \alpha_t = 0$, so that $\varphi_1^* \beta = \varphi_1^* \alpha_1 = \varphi_0^* \alpha_0 = \alpha$. Given any vector $X$ tangent to a loop $L_i$, the equality $\alpha_t(X_t, X) = \iota_{X_t} \alpha_t(X) = \gamma(X) = 0$ holds, so that the field $X_t$ is tangent to $L_i$ and the flow $\varphi_t$ preserves the $L_i$'s.

We have constructed $\varphi_1$ such that the Riemannian volume of the metric $g = \varphi_1^* g_1$ is $\varphi_1^* \text{vol}_{g_1} = \sigma$. Moreover, $\varphi_1$ is an isometry from $(M, g)$ into $(M, g_1)$ which preserves the loops $L_1, \ldots, L_q$. Since they are geodesics for $g_1$, they are also geodesics for $g$ and the result is proved. $\square$

### 2.3. Piecewise geodesic Yang-Mills measure

Let us fix on $M$ a Riemannian metric given by Proposition 2.3. We use this metric to define a subclass $\mathcal{G}$ of the set of all graphs on $M$ such that the projective limit described in Section 2.1 can be constructed when $\Lambda = \mathcal{G}$: we say that a graph $\Gamma$ belongs to $\mathcal{G}$ if its edges are piecewise geodesic and if the loops $L_1, \ldots, L_q$ belong to $\Gamma^*$. The main property of the class $\mathcal{G}$ is the following.

PROPOSITION 2.5. *Given two graphs $\Gamma_1, \Gamma_2$ in $\mathcal{G}$, there exists $\Gamma_3$ in $\mathcal{G}$ such that $\Gamma_1 \leq \Gamma_3$ and $\Gamma_2 \leq \Gamma_3$.*

Before proving this proposition, let us recall the classical property of geodesics which is the most important for our purpose. A proof of a local version of this theorem can be found for example in [19] (Proposition 3.4.2). The compactness of $M$ allows us to globalize the result.

THEOREM 2.6. *There exists a positive real number $R_M$, called convexity radius of $M$, such that if $x$ and $y$ are two points of $M$ contained in a ball of radius smaller than $R_M$, then they are joined by a unique piece of minimizing geodesic and this piece of geodesic stays inside the ball.*

This theorem implies in particular the following rigidity result:

PROPOSITION 2.7. *Let $\zeta_1$ and $\zeta_2$ be two pieces of geodesics with finite lengths. Then $\zeta_1$ and $\zeta_2$ can meet each other along at most two segments of positive length, and at isolated points for a finite number of values of their parameter.*

PROOF. Let us choose a unit-speed parametrization of the paths $\zeta_1$ and $\zeta_2$. If they meet at an infinity of points, it is easy to check that there exists a couple $(t_1, t_2)$ of times such that $\zeta_1(t_1) = \zeta_2(t_2)$ and $\dot{\zeta}_1(t_1) = \dot{\zeta}_2(t_2)$, so they are in fact two pieces of the same infinite geodesic. If this geodesic is periodic, then $\zeta_1$ and $\zeta_2$ can intersect along one or two segments. Otherwise, they have one segment in common plus a finite number of isolated points. □

PROOF OF PROPOSITION 2.5. The natural idea is to superpose $\Gamma_1$ and $\Gamma_2$. Given an edge $e$ of $\Gamma_2$, we know that $e([0,1]) \cap Supp(\Gamma_1)$ is a finite union of segments and points. So, it is possible to add a finite number of new vertices and new edges to $\Gamma_1$ in such a fashion that $e$ becomes a path in the new graph. Repeating this procedure for each edge of $\Gamma_2$ gives the result. □

Let us fix an element $(x_1, \ldots, x_q)$ of $G^q$. To each graph $\Gamma \in \mathcal{G}$, we have associated a probability space $(G^\Gamma, \mathcal{B}(G^\Gamma), P^\Gamma_{(x_1,\ldots,x_q)})$. Proposition 2.5 and the subdivision invariance property ensure that the family $(G^\Gamma, \mathcal{B}(G^\Gamma), P^\Gamma_{(x_1,\ldots,x_q)})$ of probability spaces together with the measurable maps $f_{\Gamma_1 \Gamma_2}, \Gamma_1, \Gamma_2 \in \mathcal{G}$ is a projective family. Since each $G^\Gamma$ is a compact space, Theorem 2.2 allows us to make the first step towards the construction of the Yang-Mills measure, by defining the space $(\Omega, \mathcal{A}, P_{(x_1,\ldots,x_q)})$ as the projective limit of this family. In this case, each map $f_{\Gamma_1 \Gamma_2}$ is surjective, so that each map $f_\Gamma : \Omega \longrightarrow \Omega_\Gamma$ is itself surjective.

DEFINITION 2.8. The measure $P_{(x_1,\ldots,x_q)}$ is called the *piecewise geodesic Yang-Mills measure* with respect to the metric chosen on $M$. For each path $\zeta \in \Gamma^*$, the function $h_\zeta : G^\Gamma \longrightarrow G$ gives rise by composition to a random variable defined everywhere on $\Omega$:

$$H_\zeta : \Omega \xrightarrow{f_\Gamma} G^\Gamma \xrightarrow{h_\zeta} G,$$

which is called the *random holonomy* along $\zeta$.

By definition, this limit is consistent with the discrete theory in the sense that, for each $\Gamma \in \mathcal{G}$, the natural map $f_\Gamma : \Omega \longrightarrow \Omega_\Gamma$ satisfies $(f_\Gamma)_* P_{(x_1,\ldots,x_q)} = P^\Gamma_{(x_1,\ldots,x_q)}$, so that the distribution of a family $(h_{\zeta_1}, \ldots, h_{\zeta_n})$ computed in any graph of $\mathcal{G}$ is always equal to that of $(H_{\zeta_1}, \ldots, H_{\zeta_n})$ under $P_{(x_1,\ldots,x_n)}$. Of course, the multiplicativity property is preserved.

PROPOSITION 2.9. *Let $\zeta_1$ and $\zeta_2$ be two piecewise geodesic paths such that $\zeta_1(1) = \zeta_2(0)$. Then $H_{\zeta_1 \zeta_2} = H_{\zeta_2} H_{\zeta_1}$ everywhere.*

From now on, we will use greek letters to denote piecewise geodesic paths and denote by $\Pi M$ the set of these paths.

Note that $\Omega$ can be canonically identified with the set $\mathcal{M}(\Pi M, G)$ of maps $f : \Pi M \longrightarrow G$ which satisfy $f(\zeta_1 \zeta_2) = f(\zeta_2) f(\zeta_1)$ whenever this makes sense. Using the classical Daniell-Kolmogorov theorem instead of a projective limit would have provided us with a probability measure on the set of all functions from $\Pi M$ to $G$ such that for all $\zeta_1$ and $\zeta_2$ such that $\zeta_1 \zeta_2$ exists, one has $f(\zeta_1 \zeta_2) = f(\zeta_2) f(\zeta_1)$ for almost all $f$. This is a weaker property because the exceptional set depends on $\zeta_1$ and $\zeta_2$.

At the end of this chapter, we will take a second projective limit of probability spaces in order to get a pathwise multiplicative version of the random holonomy process.

## 2.4. Lassos and small piecewise geodesic loops

Our next and main task is to prove that we can extend the family of random variables $(H_\zeta)_{\zeta \in \Pi M}$ to the class of piecewise embedded paths on $M$. To do this, we are going to approximate a path $c$ by a sequence of piecewise geodesic paths $(\zeta_n)$ and show that the sequence of random variables $(H_{\zeta_n})$ converges to a random variable $H_c$ which depends only on $c$. This will require some effort and will be achieved in several steps. In this section, we introduce some of the tools needed for the approximation procedure.

### 2.4.1. Lassos.

DEFINITION 2.10. A *lasso* is either a simple loop or a path of the form $l = cbc^{-1}$, where $c$ is an injective path and $b$ a simple loop which meets $c$ only at its base point. The loop $b$ is determined by $l$ and is called the *buckle* of the lasso $l$. If the lasso is a simple loop, it is its own buckle.

These lassos are quite different from the *lasso forms* introduced by L. Gross in [26] and studied for example by B. Driver in [20] and by L. Gross, C. King and A. Sengupta in [28].

Lassos are useful at least for two reasons: the first one is that it is easy to compute the distribution of their holonomy and the second one is that any reasonable path can be decomposed in some sense into a product of lassos. Let us begin with this second point and define what we mean by a reasonable path.

DEFINITION 2.11. A path $c \in PM$ is said to have finite self-intersection if there exists a graph $\Gamma$ such that $c \in \Gamma^*$.

This definition is not the most naive one of finite self-intersection. Indeed, it allows our path to have not only a finite number of points but also a finite number of segments as auto-intersection set. In particular, Proposition 2.7 shows that any piecewise geodesic path, which is a concatenation of injective pieces of different geodesics, has finite self-intersection.

It would be too optimistic to expect each loop with finite self-intersection to be equal to a concatenation of lassos, but to make this into a true statement, we need only replace the equality of loops by the correct equivalence relation.

DEFINITION 2.12. Two paths are said to be *basically equivalent* if one of them can be written $c_1 c_2 c_2^{-1} c_3$ and the other one $c_1 c_3$, where $c_1, c_2, c_3 \in PM$. Two paths $c$ and $c'$ are *equivalent*, and we denote $c \simeq c'$, if there exists a finite chain $c = c_0, \ldots, c_n = c'$ such that any two successive terms of this chain are basically equivalent.

This equivalence relation is natural with respect to the random holonomy, in the following sense.

LEMMA 2.13. *Let $\zeta_1$ and $\zeta_2$ be two paths of $\Pi M$ such that $\zeta_1 \simeq \zeta_2$. Then $H_{\zeta_1} = H_{\zeta_2}$ everywhere.*

PROOF. One needs only remark that if two paths of $\Pi M$ are basically equivalent, then they are basically equivalent in $\Pi M$, in that one can choose the path $c_2$ in Definition 2.12 to be piecewise geodesic. Then the result follows as a direct consequence of the multiplicativity property (Proposition 2.9). □

We can now state the decomposition result, which actually also gives some control on the length of the paths involved. The Riemannian metric on $M$ allows us indeed to compute the length of any path $c$, which we denote by $\ell(c)$. A glance at Figure 1 might make the idea easier to catch.

PROPOSITION 2.14. *Let $c$ be a path with finite self-intersection.*

*1. If $c(0) \neq c(1)$ then $c$ is equivalent to a unique product $c \simeq l_1 \ldots l_p c'$, where the $l_i$'s are lassos which are not equivalent to a constant loop and $c'$ is an injective path joining $c(0)$ to $c(1)$. Moreover, if $b_i$ denotes the buckle of the lasso $l_i$ for each $i$, the following inequality holds:*

$$\ell(c) \geq \sum_{i=1}^{p} \ell(b_i) + \ell(c'). \tag{2.2}$$

*2. If $c(0) = c(1)$, the result remains true after removing $c'$ from all expressions.*

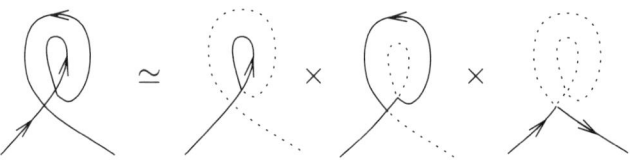

FIGURE 1. An example of lasso decomposition.

PROOF. We proceed by induction on the number of edges in a decomposition of $c$ as a path in a graph. If $c$ is an edge, we are in the first case and the result is true. Suppose that $c = e_1 \ldots e_r$. If $c$ is an injective path or a simple loop, the result is true. Otherwise, let us consider the first time at which $c$ intersects itself and proceed by induction. Let $i$ be the smallest integer such that there exists $1 \leq j < i$ with $e_j(0) = e_i(1)$. Such an $i$ exists, and $(i, j) \neq (r, 1)$. We have:

$$c \simeq e_1 \ldots e_{j-1} \cdot e_j \ldots e_i \cdot (e_1 \ldots e_{j-1})^{-1} \cdot e_1 \ldots e_{j-1} \cdot e_{i+1} \ldots e_r.$$

It can happen that the either first piece or the last piece of this decomposition are empty, respectively if $j = 1$ or $i = r$, but not simultaneously. If $j = i - 1$, it is possible that $e_i = e_j^{-1}$ so that $e_j e_i$ is equivalent to a constant path. This cannot happen if $j < i - 1$, in which case $e_j \ldots e_i$ is a genuine simple path. Thus, the product of the three first terms is either equivalent to a constant path, or is a simple loop (if $j = 1$), or a lasso. The product of the two last terms is the product of a number of edges which is positive and strictly less than $p$. So by induction, this path $\tilde{c}$ is equivalent to $l_1 \ldots l_q c'$, or to $l_1 \ldots l_q$ if $c$ is a loop. Note that $\ell(c) = \ell(e_1) + \ldots + \ell(e_r) = \ell(\tilde{c}) + \ell(e_j) + \ldots + \ell(e_i)$. So, by induction, $\ell(c) \geq \ell(b_1) + \ldots + \ell(b_q) + \ell(c') + \ell(e_j) + \ldots + \ell(e_i)$. In the case $j = i - 1$ and $e_j = e_i^{-1}$, we have $c \simeq \tilde{c}$ and the result is true with a strict inequality. Otherwise, there exists a lasso $l_0$ such that $c \simeq l_0 \tilde{c}$ and the length of the buckle of this lasso $l_0$ is exactly $\ell(e_j) + \ldots + \ell(e_i)$. □

### 2.4.2. Holonomy along small piecewise geodesic loops.

Let us consider now the distribution of the random holonomy along a lasso. Since, by multiplicativity, the holonomy along a lasso is conjugated to the holonomy along its buckle, and since the distance to the identity in $G$ is invariant by conjugation, Proposition 1.29 shows that we can control the holonomy along a lasso if we control the area enclosed by its buckle. This suggests that small lassos are those whose buckle encloses a small area. Nevertheless, Proposition 2.14 involves the length of the buckles rather than the area they enclose. The link between these quantities is provided by an isoperimetric inequality.

PROPOSITION 2.15. *There exist $R > 0$ and $K_{iso} > 0$ such that any simple loop $l$ on $M$ contained in a ball $B$ of radius $R$ is the boundary of an open set $U \subset B$ such that*

$$\sigma(U) \leq K_{iso}\ell(l)^2. \tag{2.3}$$

PROOF. Take $R_0 > 0$ such that any closed geodesic ball of $M$ of radius equal to $R_0$ is diffeomorphic to a disk. Let $B_1, \ldots, B_n$ be a finite covering of $M$ by open balls of radius $R_0$. On each of these balls, the metric on $M$ can be compared to a pull-back of the Euclidean metric on the unit disk in $\mathbb{R}^2$, and the usual isoperimetric inequality on $\mathbb{R}^2$ gives rise to an isoperimetric inequality similar to (2.3). Let $K_{iso}$ be the largest constant appearing in these inequalities. We claim that the result holds with this constant, for some $R > 0$.

Indeed, it is a classical fact that, given an open covering $(O_i)_{i \in I}$ of a compact metric space, there exists $r > 0$ such that any ball of radius $r$ is contained in one of the open sets $O_i$. Taking $R$ to be such a radius corresponding to the covering $B_1, \ldots, B_n$ of $M$, we see that (2.3) holds with the constant $K_{iso}$ in any ball of radius $R$. □

Now we can estimate the holonomy along a small piecewise geodesic lasso in terms of the length of its buckle. We use the notation $E_{(x_1,\ldots,x_q)}$ or $E_x$ for the expectation with respect to $P_{(x_1,\ldots,x_q)}$.

PROPOSITION 2.16. *There exist $L_0 > 0$ and $K > 0$ such that for any piecewise geodesic lasso $\lambda$ with buckle $\beta$ such that $\ell(\beta) < L_0$,*

$$E_x\rho(H_\lambda) = E_x d(1, H_\lambda) \leq K\ell(\beta).$$

PROOF. By definition of the buckle of a lasso, $\lambda$ can be written $\zeta\beta\zeta^{-1}$ for some path $\zeta$, and, by the multiplicativity of the holonomy and the invariance by conjugation of the distance on $G$, we have $E_x\rho(H_\lambda) = E_x\rho(H_\beta)$. Let $L_0$ be shorter than the shortest length of a loop non homotopic to a point on $M$ and also shorter than the radius $R$ given by Proposition 2.15. Then the assumption $\ell(\beta) < L_0$ implies that $\beta$ is the boundary of a subset $D$ of $M$ diffeomorphic to a disk. By using Proposition 1.29, we get

$$E_x\rho(H_\beta) \leq C\sqrt{\sigma(D)} \leq C\sqrt{K_{iso}}\,\ell(\beta)$$

hence the result. □

This result suggests the use of the following distance between $G$-valued random variables.

DEFINITION 2.17. Let $X$ and $Y$ be two $G$-valued random variables defined on the same probability space. The distance $d_P(X,Y)$ is defined by:
$$d_P(X,Y) = Ed(X,Y),$$
where $d$ is the unit-volume bi-invariant Riemannian distance on $G$.

Since $G$ is compact, hence bounded, this distance induces the topology of the convergence in probability. The probability we are considering on $\Omega$ is of course $P_{(x_1,\ldots,x_q)}$, where $(x_1,\ldots,x_q)$ is still fixed.

We could reformulate Proposition 2.16 using the distance $d_P$, but we can actually get a much more powerful result by putting it together with the decomposition result of Proposition 2.14.

PROPOSITION 2.18. *For any piecewise geodesic loop $\zeta$ such that $\ell(\zeta) < L_0$, we have*
$$d_P(1, H_\zeta) \leq K\ell(\zeta),$$
where $K$ and $L_0$ are the constants defined by Proposition 2.16.

PROOF. Since $\zeta$ is piecewise geodesic, it has finite self-intersection, so that it is equivalent to a product of piecewise geodesic lassos, say $\zeta \simeq \lambda_1 \ldots \lambda_p$. This gives:

$$\begin{aligned}
d_P(1, H_\zeta) &= d_P(1, H_{\lambda_1} \ldots H_{\lambda_p}) \\
&\leq d_P(1, H_{\lambda_1}) + \ldots + d_P(H_{\lambda_1} \ldots H_{\lambda_{p-1}}, H_{\lambda_1} \ldots H_{\lambda_p}) \\
&\leq d_P(1, H_{\lambda_1}) + \ldots + d_P(1, H_{\lambda_p}) \\
&\leq K(\ell(\beta_1) + \ldots + \ell(\beta_p)) \\
&\leq K\ell(\zeta),
\end{aligned}$$

where we have used successively the triangular inequality, the left invariance of the distance on $G$, Proposition 2.16 and inequality (2.2). □

**2.4.3. Double layer potential of small piecewise geodesic loops.** It happens that the techniques which we have just used to estimate the holonomy along a small loop work just as well to estimate the double layer potential of a small loop (recall Definition 1.39). This is the first step in the proof of Proposition 2.45, which will play an important role in the study of the continuous Abelian theory.

We keep the Riemannian metric on $M$, but we do not need to consider piecewise geodesic paths in this paragraph, since we do not compute any random holonomy.

PROPOSITION 2.19. *Let $l$ be a lasso with buckle $b$ such that $\ell(b) < L_0$, where $L_0$ is the length defined by Proposition 2.16. Then,*
$$\| u_l \|_{L^2} \leq \sqrt{K_{iso}}\, \ell(b).$$

PROOF. As in the proof of Proposition 2.16, we see that $b$ is necessarily the boundary of a disk $D$ such that $\sigma(D) \leq K\ell(b)^2$. Thus, by Proposition 1.40,
$$\| u_b \|_{L^2} = \left( \frac{\sigma(D)\sigma(D^c)}{\sigma(M)} \right)^{\frac{1}{2}} \leq \sigma(D)^{\frac{1}{2}} \leq \sqrt{K_{iso}}\, \ell(b).$$

By additivity of the double layer potential, we have $u_l = u_b$ almost everywhere, hence the result. □

As in the preceding paragraph, we can extend this result to loops with finite self-intersection.

PROPOSITION 2.20. *Let $l$ be a loop with finite self-intersection such that $\ell(l) < L_0$. Then*
$$\| u_l \|_{L^2} \leq \sqrt{K_{iso}}\, \ell(l).$$

PROOF. The loop $l$ is equivalent to a product of lassos, say $l \simeq l_1 \ldots l_p$, with buckles $b_1, \ldots, b_p$ respectively. Two equivalent paths have the same double layer potential almost everywhere, so that

$$\begin{aligned} \| u_l \|_{L^2} &\leq \| u_{l_1} + \ldots + u_{l_p} \|_{L^2} \\ &\leq \| u_{l_1} \|_{L^2} + \ldots + \| u_{l_p} \|_{L^2} \\ &\leq \sqrt{K_{iso}}\, (\ell(b_1) + \ldots + \ell(b_p)) \\ &\leq \sqrt{K_{iso}}\, \ell(c) \end{aligned}$$

and the result is proved. □

Propositions 2.16 and 2.18 on one hand and 2.19 and 2.20 on the other hand provide very similar estimates for the random holonomy and the double layer potential along small lassos or loops. We will use this similarity to rewrite directly more sophisticated results on the random holonomy as results on the double layer potential in Sections 2.6.3 and 2.6.6.

## 2.5. The space of paths

In this section, we summarize some technical properties of paths which will be useful in the construction of the Yang-Mills measure. Let us begin by endowing $PM$ with an appropriate topology.

### 2.5.1. Convergence of paths.

In order to approximate arbitrary paths by piecewise geodesic paths, we need an appropriate topology on $PM$. All estimates we have established so far rely, more or less directly, on a control of the area, inside a loop or between two loops. However, there is no well-defined quantity such as the area of the interstice between two paths, and we must find another sensible quantity instead. As a first attempt, we may use the uniform distance between two paths defined by

$$d_\infty(c_1, c_2) = \inf \sup_{t \in [0,1]} d(c_1(t), c_2(t)),$$

where the infimum is taken over all parametrizations of $c_1$ and $c_2$. It is true but not obvious that $d_\infty$ is a distance on $PM$. However, we do not pause to prove this fact, because we do not need it.

Indeed, this distance, which has been adopted by C. Becker in [**11**], is not fine enough for our purpose and needs to be modified. C. Becker claimed that the double layer potential of a loop of $\mathbb{R}^2$, as an element of $L^2(\mathbb{R}^2, \sigma)$, depends continuously on this loop when the set of loops on $M$ is endowed with the topology induced by the uniform distance. This is true if one restricts to simple loops, but not if one allows loops to have a self-intersection, even a finite one. Let us describe a counterexample, in order to understand how $d_\infty$ should be modified.. To work on $\mathbb{R}^2$ or on a compact surface does not change the main idea. Let $M$ be the sphere $S^2$ embedded as usual in $\mathbb{R}^3$, endowed with the standard metric. Let us consider the pencil of planes containing the horizontal line $z = 0, y = -1$. Each such plane $P$ separates two pieces on $S^2$, one above $P$ and one beneath. We index the pencil by the interval $[0, 4\pi)$ in such a way that the piece of $S^2$ beneath $P_t$, which we denote by $B_t$, has area $t$. Finally, for each $t \in [0, 4\pi)$, we consider the loop $c_t$ based at $(0, -1, 0)$ which bounds $B_t$ with the usual orientation convention, that is, clockwise when looked at from above. Let $0 < t_1 < \ldots < t_n < 4\pi$ be $n$ distinct times and set $c = c_{t_1} \ldots c_{t_n}$. For each $i$, $u_{c_i} = \mathbf{1}_{B_{t_i}} - \frac{t_i}{4\pi}$. Thus,

$$\| u_c \|_{L^2}^2 = \| \mathbf{1}_{B_{t_1}} + \ldots + \mathbf{1}_{B_{t_n}} \|_{L^2}^2 + \frac{(t_1 + \ldots + t_n)^2}{16\pi^2}$$
$$- 2 \left( \frac{t_1 + \ldots + t_n}{4\pi}, \mathbf{1}_{C_{t_1}} + \ldots + \mathbf{1}_{C_{t_n}} \right)_{L^2}$$
$$\geq n^2 t_1 + \frac{n^2 t_1^2}{16\pi^2} - \frac{n^2 t_n^2}{2\pi}.$$

Suppose that $t_1 \geq \frac{1}{n}$ and $t_n \leq \frac{2}{n}$ (take for example $t_i = \frac{n+i}{n^2}$). Then $\| u_c \|_{L^2}^2$ is of the order of $n$, so if we let $n$ tend to infinity, this norm also tends to infinity. But at the same time, the loop $c$ tends to the constant loop at $(0, -1, 0)$ in the topology induced by the distance $d_\infty$. The potential of this constant loop being equal to zero, this shows that there is no continuity.

The problem in this counterexample is that, although the loops grow closer to a single point, they wind a great number of times around small domains. To avoid this behaviour, it is necessary to endow the space of paths with a topology finer than that induced by $d_\infty$. In the previous paragraphs, the length of the paths has proved an efficient way of controlling holonomies, whence we define the following new distance.

DEFINITION 2.21. On the set of paths $PM$, we define the *distance in length* $d_\ell$ by
$$d_\ell(c_1, c_2) = \max(d_\infty(c_1, c_2), |\ell(c_1) - \ell(c_2)|).$$

In order to show that $d_\ell$ is a genuine distance, we are going to prove two lemmas. The first one is a strong statement of the semi-continuity of the length with respect to the uniform topology. The second one says that if two paths are $d_\ell$-close to each other, then their uniform distance is small when both are parametrized by arc length. That $d_\ell$ is indeed a distance follows easily from this result. Observe that, in these two lemmas, we do not use more than the fact that all paths of $PM$ have bounded variation.

LEMMA 2.22. *Let $c$ be a parametrized path. For all $\varepsilon > 0$ there exists $\delta > 0$ such that if $c'$ is another parametrized path and if $\sup_t d(c(t), c'(t)) < \delta$, then $\ell(c') > \ell(c) - \varepsilon$. Moreover, it is possible to choose $\delta$ such that the inequality $\ell(c') < \ell(c) + \delta$ implies*
$$\sup_{s<t} |\ell_{s,t}(c) - \ell_{s,t}(c')| < \varepsilon,$$
*where $\ell_{s,t}(c)$ denotes the length of the restriction of $c$ to the interval $[s, t]$.*

PROOF. The first part of the statement is classical. Let us prove it shortly and introduce some notation. If $\triangle = \{0 = t_0 < t_1 < \ldots < t_n = 1\}$ is a subdivision of $[0, 1]$, set $\ell^\triangle(c) = \sum_i d(c(t_i), c(t_{i+1}))$. Then $\ell(c) = \sup_\triangle \ell^\triangle(c)$ with the supremum taken over all subdivisions. Given $\varepsilon > 0$, let $\triangle$ be such that $\ell(c) < \ell^\triangle(c) + \varepsilon/8$, set $n = \sharp\Delta - 1$ and $\delta = \varepsilon/16n$. If $c'$ satisfies $\sup_t d(c(t), c'(t)) < \delta$, then $\ell^\Delta(c') \geq \ell^\Delta(c) - \varepsilon/8 > \ell(c) - \varepsilon/4$. In particular, $\ell(c') > \ell(c) - \varepsilon$.

Now assume that $\ell(c') < \ell(c) + \delta$. Let us fix $t$ in $[0,1]$ and set $\Delta' = \Delta \cup \{t\}$. On one hand, we have both $\ell_{0,t}(c) < \ell_{0,t}^{\Delta'}(c) + \varepsilon/8$ and $\ell_{t,1}(c) < \ell_{t,1}^{\Delta'}(c) + \varepsilon/8$, so that the argument above gives $\ell_{0,t}(c') > \ell_{0,t}(c) - \varepsilon/4$ and $\ell_{t,1}(c') > \ell_{t,1}(c) - \varepsilon/4$. On the other hand,

$$\ell_{0,t}(c') < \ell_{0,t}(c) + \ell_{t,1}(c) - \ell_{t,1}(c') + \delta < \ell_{0,t}(c) + \varepsilon/2,$$

because $\delta < \varepsilon/4$. Finally, we find $|\ell_{0,t}(c') - \ell_{0,t}(c)| < \varepsilon/2$ and hence the result. □

LEMMA 2.23. *Let $c$ be an element of $PM$. For all $\varepsilon > 0$, there exists $\eta > 0$ such that if $c'$ is another path of $PM$ and if $d_\ell(c, c') < \eta$, then, with the arc-length parameter for both $c$ and $c'$, one has*

$$\sup_t d(c(t), c'(t)) < \varepsilon.$$

PROOF. Choose $c$ in $PM$ and $\varepsilon > 0$. Let $\delta$ be given by Lemma 2.22 applied to $c$ with its arc-length parametrization and $\varepsilon/6$. Set $\eta = \min(\varepsilon/4, \delta, \ell(c)/2)$ and take $c'$ such that $d_\ell(c, c') < \eta$. Fix the arc-length parametrization for $c'$.

Since $d_\infty(c, c') < \varepsilon/4$, there exists a reparametrization $\phi$, that is, a continuous increasing piecewise diffeomorphism of $[0,1]$ such that $\sup_t d(c(t), c'(\phi(t))) < \varepsilon/4$. This implies

$$\sup_t d(c(t), c'(t)) \leq \varepsilon/4 + \sup_t d(c'(t), c'(\phi(t))) = \varepsilon/4 + \ell(c') \sup_t |t - \phi(t)|.$$

By Lemma 2.22 applied to the parametrized paths $c$ and $c' \circ \phi$, we have $\sup_t |\ell(c)t - \ell(c')\phi(t)| < \varepsilon/6$. This implies

$$\ell(c') \sup_t |t - \phi(t)| < \frac{\ell(c')}{\ell(c)} \left( \frac{\varepsilon}{6} + |\ell(c) - \ell(c')| \right).$$

Since $|\ell(c) - \ell(c')| < \ell(c)/2$ by definition of $\eta$, we have $\ell(c')/\ell(c) < 3/2$ and the result follows easily. □

REMARK 2.24. This lemma implies immediately that $d_\ell$ is a distance on $PM$. Moreover, the topology induced by $d_\ell$ is not changed if one replaces the distance $d$ on $M$ by another equivalent distance. Hence, the topology induced by $d_\ell$ does not depend on the choice of the Riemannian metric on $M$. We call *convergence in length* the notion of convergence associated with this intrinsic topology.

REMARK 2.25. Let $V(c_1, c_2)$ be defined for two paths $c_1$ and $c_2$ of $PM$ by

$$V(c_1, c_2) = \inf_{\text{param.}} \sup_\Delta \sum_{j=1}^n |d(c_1(t_{j-1}), c_1(t_j)) - d(c_2(t_{j-1}), c_2(t_j))|,$$

where the supremum is taken over all subdivisions of the interval $[0,1]$. It can be shown that convergence in length is equivalent on $PM$ to the a priori stronger convergence defined by the distance

$$d_1(c_1, c_2) = \max(d_\infty(c_1, c_2), V(c_1, c_2)),$$

although $d_\ell$ and $d_1$ are not equivalent. Of course, this depends on the fact that the paths of $PM$ have more than just bounded variation. General results concerning convergence in length and in variation can be found in [**34, 1**].

The convergence in length is compatible with the concatenation of paths in the sense of the following obvious result.

PROPOSITION 2.26. *If $c$, $c'$ are two paths such that $cc'$ exists and if $(c_n)_{n\geq 0}$, $(c'_n)_{n\geq 0}$ are two sequences of paths such that $c_n$ converges in length to $c$ and $c'_n$ to $c'$, and if for all $n$ the path $c_n c'_n$ exists, then $c_n c'_n$ converges in length to $cc'$.*

**2.5.2. Tubular neighbourhoods and Fermi coordinates.** We continue with a strong geometrical property of embedded paths in the case where $M$ has no boundary. Recall that any element of $PM$ is the concatenation of a finite number of such embedded paths.

Let us fix a path $c$ which is an embedded submanifold and choose a parametrization of $c$. In this paragraph, we assume that $c$ is parametrized on an interval $I$ containing $[0, 1]$, but which can be open. Let us choose a continuous unit vector field $N$ along $c$, normal to $c$. Recall that, given a point $m$ of $M$ and a vector $X$ tangent to $M$ at $m$, $\exp_m(X)$ is by definition the endpoint of the geodesic segment of length $\|X\|$ starting at $m$ in the direction $X$. A proof of the following result can be found in [**25**].

PROPOSITION 2.27. *There exists a positive real number $r$ such that the mapping*

$$\psi : I \times [-r, r] \longrightarrow M$$
$$(t, s) \longmapsto \exp_{c(t)}(s N_{c(t)})$$

*is a diffeomorphism onto its image, which is called* tubular neighbourhood *of $c$ or* a tube *around $c$. The coordinates $(t, s)$ are called* Fermi coordinates. *They satisfy the following properties:*

1. *For any fixed $t_0$, the curve $s \mapsto \psi(t_0, s)$ is a piece of geodesic normal to $c$.*
2. *For any couple $(t, s) \in I \times [-r, r]$, $d(\psi(t, s), c(I)) = s$.*
3. *The vector fields $\psi_* \partial_t$ and $\psi_* \partial_s$ are orthogonal.*

Properties 2 and 3 give us a good control on the geometry of $M$ in a neighbourhood of the range of $c$. Property 3 is a generalization of the Gauss lemma. We shall always take the radius of any tubular neighbourhood smaller than the convexity radius $R_M$ of $M$, defined in Theorem 2.6.

When $M$ has a boundary, this result fails: for example, if $N$ is a component of the boundary of $M$, then $N$ is an embedded submanifold which has no tubular neighbourhood. However, by embedding $M$ in a closed manifold, we see that $N$ has a half of a tubular neighbourhood.

Let us give an easy consequence of this fact. If $r$ denotes the radius of a half-tube around $N$, then the subset of $M$ defined by the equation $0 \leq s \leq r/2$ in Fermi coordinates is a closed neighbourhood of $N$ of which $N$ is a strong deformation retract. We have used this property in the proof of Proposition 2.3.

Observe that, when $M$ has a boundary, a smooth path does not always have even a half of a tubular neighbourhood.

On a closed surface, the existence of tubular neighbourhoods allows us to prove the following technical result.

LEMMA 2.28. *Let $c$ be an embedded path. There exists $\varepsilon > 0$ such that, for all $R < \varepsilon$ and all $t \in [0, 1]$, $c^{-1}(B(c(t), R))$ is an interval.*

PROOF. By definition of a path, there exist $\eta > 0$ and an embedding $\bar{c} : (-\eta, 1 + \eta) \longrightarrow M$ such that $c$ is the restriction to $[0, 1]$ of $\bar{c}$. Let $\bar{T}$ be a tubular neighbourhood around $\bar{c}$ of radius smaller than $r$. We claim that the lemma holds with $\varepsilon = d(c([0,1]), M - \overline{T})$.

For, let $R$ be smaller than $\varepsilon$, choose $t$ in $[0,1]$ and consider the sphere $S(c(t), R)$. It is contained in $\overline{T}$ by definition of $\varepsilon$. Our claim is equivalent to the fact that this sphere cuts $\bar{c}$, hence $c$, at most twice. Suppose it cuts $\bar{c}$ at least three times. Then the curve $S(c(t), R)$ has at least three points where its tangent is horizontal in Fermi coordinates, i.e. with a tangent parallel to $\partial_t$. Since the sphere has exactly two points on the geodesic normal to $c$ at $c(t)$, namely $\psi(t, R)$ and $\psi(t, -R)$, there is at least one point, say $p$, of $S(c(t), R)$ where the tangent is horizontal and whose first Fermi coordinate is different from $t$. Let us consider the geodesic radius from $c(t)$ to $p$. By the Gauss lemma, it is orthogonal to the sphere at $p$ and since $\partial_t$ and $\partial_s$ are orthogonal, it is in fact a vertical line in Fermi coordinates around $\bar{c}$. This implies that $p$ and $c(t)$ have the same first Fermi coordinate, which contradicts our choice of $p$. $\square$

### 2.5.3. Approximation by piecewise geodesic paths.

In this paragraph, we keep assuming that $M$ has no boundary. The main approximation result says that each embedded path can be nicely approximated by a piecewise geodesic path satisfying some technical conditions which will allow us to estimate easily its holonomy.

Let us consider an embedded path $c$ with a tubular neighbourhood of radius $r$.

**PROPOSITION 2.29.** *Let $x, y, z$ be such that $0 \leq x < y < z \leq r$. There exists a piecewise geodesic path $\sigma$ such that*

1. $\sigma(0) = \psi(0, y)$ *and* $\sigma(1) = \psi(1, y)$,
2. $\sigma((0,1)) \subset \psi((0,1) \times (x, z))$,
3. $\sigma$ *is injective. Moreover, if $c_y$ denotes the path $t \mapsto \psi(t, y)$, then $\sigma$ can be chosen such that its length is arbitrarily close to that of $c_y$.*

The path $c_y : t \mapsto \psi(t, y)$ satisfies properties 1. to 3., but of course it is not piecewise geodesic in general. Still, we are going to construct $\sigma$ by approximating $c_y$, as one would approximate a curve in $\mathbb{R}^2$ by a piecewise linear path.

PROOF OF PROPOSITION 2.29. By definition of a path, $c$ is the restriction to $[0,1]$ of some embedding $\bar{c} : [-\varepsilon, 1+\varepsilon] \longrightarrow M$. Let us consider a tube $\overline{T}$ of radius $r$ around $\bar{c}$ and set $\bar{c}_y : t \mapsto \psi(t, y)$ in this larger tube. Let us denote by $c_y$ the restriction of $\bar{c}_y$ to $[0,1]$. Since the speed of $c_y$ is bounded on $[0,1]$, the quantity

$$\delta_n(c_y) = \sup_{1 \leq k \leq n} d\left(c_y\left(\frac{k-1}{n}\right), c_y\left(\frac{k}{n}\right)\right)$$

tends to 0 as $n$ tends to infinity. Hence, for $n$ large enough, it is smaller than the convexity radius $R_M$, than $\min(z-y, y-x)$ and than the distance between $c_y([0,1])$ and $M - \overline{T}$. For such an $n$, and for all $k = 1, \ldots, n$, $\bar{c}_y(\frac{k-1}{n})$ and $c_y(\frac{k}{n})$ are close enough to be joined by a unique minimizing geodesic, which we denote by $\zeta_{n,k}$. This geodesic stays inside the geodesic ball of radius $\delta_n$ around $c_y(\frac{k}{n})$, hence it stays inside the tube around $\bar{c}$.

We claim that $\zeta_n$ defined by $\zeta_n = \zeta_{n,1} \ldots \zeta_{n,n}$ is the graph of a continuous function in Fermi coordinates. More precisely, there exists a continuous function $u_n : [0,1] \longrightarrow (-r, r)$ such that for each $t \in [0,1]$, $\zeta_n(t) = \psi(t, u_n(t))$. Observe at this point that, by taking $n$ larger is necessary, we can ensure that the length of $\zeta_n$ is arbitrarily close to that of $c_y$.

We have proved that $\zeta_{n,k}$ stays inside the tube. Each vertical slice $t=t_0$ of the tube is a minimizing piece of a geodesic which meets $c$ only once inside the tube, so that it meets $\zeta_{n,k}$ at most one time. Thus, $\zeta_{n,k}$ is the graph of a continuous function defined on the segment $[\frac{k-1}{n},\frac{k}{n}]$, equal to $y$ at both endpoints of this segment. We can put these functions together and construct $u_n$, which is continuous.

As the graph of a function, $\zeta_n$ is necessarily injective, so it satisfies property 3. The inequality $|u_n(t)-y|=d(\zeta_n(t),c_y)\leq\delta_n$ shows that $\zeta_n$ stays in $\psi([0,1]\times(x,z))$. Together with the fact that $\zeta_n$ meets at most once each vertical boundary, this implies property 2. Since property 1. is a direct consequence of the definition of $\zeta_n$, $\sigma=\zeta_n$ has all the required properties. □

This result allows us to construct a whole sequence of approximating paths. For each $n\geq 0$, let us put

$$x_n=\frac{r}{2^{n+1}},\quad y_n=\frac{3}{2}\frac{r}{2^{n+1}},\quad z_n=\frac{r}{2^n}$$

and consider a path $\sigma_n$ defined by Proposition 2.29, such that $|\ell(\sigma_n)-\ell(c_{y_n})|<1/n$. This path does not have the same endpoints as $c$, and this would be a problem in the sequel. So, let us define $\lambda_n$ as the vertical segment joining $\psi(0,0)$ to $\psi(0,y_n)$ and $\rho_n$ as the vertical segment joining $\psi(1,y_n)$ to $\psi(1,0)$. The paths $\alpha_n=\lambda_n\sigma_n\rho_n$ are our candidates for a convergence result for the random holonomy. For the moment, let us check that they approximate $c$ as we expect them to do.

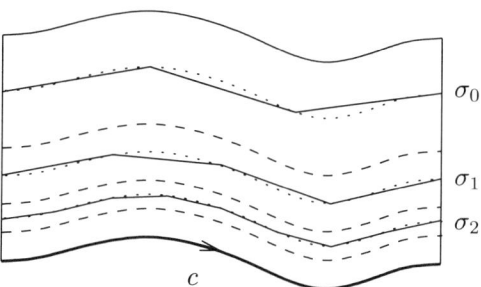

FIGURE 2. The upper half of the tube around $c$ and the first terms of the sequence $(\sigma_n)$.

LEMMA 2.30. *The sequence $(\alpha_n)_{n\geq 0}$ converges in length to $c$ with fixed endpoints.*

PROOF. It is clear that $d_\infty(\alpha_n,c)$ tends to 0 and the statement $\ell(\alpha_n)\to\ell(c)$ is equivalent to $\ell(c_y)\to\ell(c)$ as $y$ tends to 0, which is also clearly true. □

**2.5.4. Surfaces with boundary.** Recall that $\Pi M\subset PM$ denotes the set of piecewise geodesic paths. In the preceding paragraph, we have established that $\Pi M$ is dense in $PM$ for the topology of the convergence in length when $M$ has no boundary. We want to prove that this is true in general.

When $M$ has a boundary, let $\overset{\circ}{PM}$ denote the set of paths whose image is contained in the interior of $M$. Any embedded path of $\overset{\circ}{PM}$ has a tubular neighbourhood in $M$ and can be approximated by piecewise geodesic paths. Hence it is sufficient to prove the following lemma.

LEMMA 2.31. $\overset{\circ}{PM}$ *is dense in* $PM$.

PROOF. To prove this, we choose a convenient metric on $M$, one such that each component of $\partial M$ has a neighbourhood diffeomorphic to $S^1 \times [0,1) \ni (\theta, t)$ in which the metric is $d\theta^2 + dt^2$. Let $f : [0, 1) \longrightarrow [0, 1]$ be a smooth function such that $f(t) = 1$ if $t \leq 1/2$ and $f(t) = 0$ if $t \geq 3/4$. Let $\varphi_t$ be the flow at time $t$ of the vector field $f(t)\partial_t$. For $t \leq 1/4$, $\varphi_t$ sends isometrically $S^1 \times [0, 1/4]$ onto $S^1 \times [t, t+1/4]$.

If $c$ is a path contained in the neighbourhood of a component of $\partial M$ corresponding to $S^1 \times [0, 1/4]$, it is approximated in length by $\varphi_t \circ c$ as $t$ tends to 0. Better, for $t$ small enough, we can concatenate $\varphi_t \circ c$ with short geodesics to make it have the same endpoints as $c$, without altering the convergence in length.

Finally, any path can be written as a finite products of paths contained in $S^1 \times [0, 1/4]$ and paths remaining outside $S^1 \times [0, 1/8]$. By compatibility of the concatenation with the convergence in length, this implies the result. □

We finish this section with a density result *with fixed endpoints*. This restriction will indeed play an important role in the construction of the random holonomy.

PROPOSITION 2.32. *Whether $M$ has a boundary or not, every path of $PM$ can be approximated in length with an arbitrary precision by a piecewise geodesic path having the same endpoints.*

PROOF. When $M$ is closed, this follows immediately from Lemma 2.30, because each $\alpha_n$ has the same endpoints as $c$.

If $M$ has a boundary, we need to be more careful, because it is not true that every point of the boundary has a geodesically convex neighbourhood. So, if $c$ is a path on $M$ with an endpoint on $\partial M$, we begin by concatenating it with a short geodesic normal to the boundary in order to make both endpoints of $c$ belong to the interior of $M$. Then we can take a sequence of piecewise geodesic paths approximating this slightly modified path and concatenate the terms of this sequence with short geodesics to ensure that they converge with fixed endpoints. If the normal segment added to $c$ is short enough and if we take a good enough approximation of this modified path, we get a good approximation of $c$. □

## 2.6. Definition of the random holonomy

Until Section 2.9 and unless explicitly stated, we keep assuming that $M$ has no boundary. Let us fix an embedded path $c$. In Section 2.5.3, we have defined a nice sequence $(\alpha_n)$ of piecewise geodesic paths which approximate $c$. We need now to prove that the random holonomies along these paths converge in the topology induced by the distance $d_P$, and that the limit depends only on the path $c$.

In order to be able to use our estimates, we make the assumption that the area of the tubular neighbourhood of $c$ is smaller than the area $s$ defined by Proposition 1.29.

### 2.6.1. Existence of a limit random holonomy along embedded paths.

PROPOSITION 2.33. *The sequence of random variables $(H_{\alpha_n})_{n \geq 0}$ defined on the probability space $(\Omega, \mathcal{A}, P_{(x_1,\ldots,x_q)})$ is a Cauchy sequence with respect to the distance $d_P$.*

PROOF. Let $m \geq n$ be two integers. We want to estimate $d_P(H_{\alpha_m}, H_{\alpha_n}) = d_P(1, H_{\alpha_n^{-1}\alpha_m})$. The properties of the paths $\sigma_n$ and $\sigma_m$ show that $\alpha_n^{-1}\alpha_m$ is equivalent to a simple loop which is the boundary of an open set contained in $\psi([0,1] \times [0, \frac{r}{2^n}])$. Thus, the assumption on the area of the tube allows us to apply proposition 1.29. We get:

$$d_P(H_{\alpha_m}, H_{\alpha_n}) \leq C\sigma\left(\psi\left([0,1] \times \left[0, \frac{r}{2^n}\right]\right)\right)^{\frac{1}{2}} \leq C' 2^{-\frac{n}{2}}$$

and the result is proved. □

Since $G$ is a bounded complete metric space, the space of $G$-valued random variables on $(\Omega, \mathcal{A}, P_{(x_1,\ldots,x_q)})$ endowed with the distance $d_P$ is also complete.

In particular, the sequence $(H_{\alpha_n})_{n \geq 0}$ has a limit which we denote by $H_c$, anticipating the fact that it depends only on $c$.

REMARK 2.34. The relation $H_{\alpha_n} \xrightarrow{P} H_c$ defines a random variable $H_c$ only $P_{(x_1,\ldots,x_q)}$-almost surely. If we change the values of $x_1, \ldots, x_q$, then Proposition 2.33 produces some other random variable $H'_c$, but nothing tells us that $H_c = H'_c$, even almost surely for some measure. For this reason, we will keep $x = (x_1, \ldots, x_q)$ fixed during the construction of the random holonomy along arbitrary paths and show in the end that this problem of non-unicity of the variable associated to a path can be solved. In fact, it should be kept in mind that we are constructing finite-dimensional marginals of a process rather than random variables.

### 2.6.2. Unicity of the limit random holonomy along embedded paths.

The next step is indeed to check that any other approximating sequence of piecewise geodesic paths would have given rise to the same random variable. This will be a consequence of the following continuity property.

PROPOSITION 2.35. *For all $\varepsilon > 0$, there exists $\delta > 0$ such that if $\alpha$ is a piecewise geodesic path with the same endpoints as $c$ and such that $d_\ell(c, \alpha) < \delta$, then $d_P(H_c, H_\alpha) < \varepsilon$.*

Let us emphasize that, in this paragraph, $H_c$ denotes the limit in probability of the particular sequence $(H_{\alpha_n})_{n \geq 0}$ defined previously. The path $c$ is the same as in the preceding section, and it still has a tubular neighbourhood with Fermi coordinates (see Proposition 2.27). We begin by proving a slightly weaker statement.

LEMMA 2.36. *For all $\varepsilon > 0$, there exists $\delta > 0$ such that if $\alpha$ is an injective piecewise geodesic path with the same endpoints as $c$, such that $d_\infty(c, \alpha) < \delta$ and $\alpha((0,1)) \subset \psi((0,1) \times (-r, r))$, then $d_P(H_c, H_\alpha) < \varepsilon$.*

It is not necessary to control $|\ell(\alpha) - \ell(c)|$ here because $\alpha$ is injective, and cannot wind wildly around domains of $M$ of small area.

PROOF. Let $C$ be the constant given by Proposition 1.29. Let $n$ be such that $d_P(H_c, H_{\alpha_n}) < \varepsilon/2$ and $C\sigma(\psi([0,1] \times [-\frac{r}{2^n}, \frac{r}{2^n}]))^{1/2} < \varepsilon/2$. Set $\delta = 2^{-(n+1)}$ and

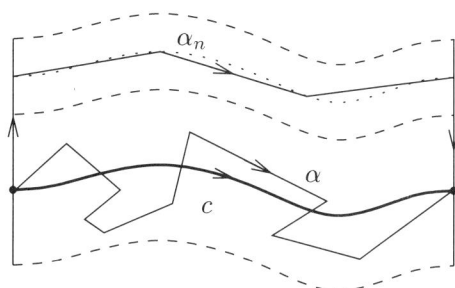

FIGURE 3.

suppose that $d_\infty(c,\alpha) < \delta$. Then the range of $\alpha$ is contained in the strip $|s| < 2^{-(n+1)}$ in Fermi coordinates, so that $\alpha$ meets $\alpha_n$ only on the 'vertical' boundary of the tube. By assumption, $\alpha$ has only its endpoints on this boundary, hence it meets $\alpha_n$ only at its own endpoints, which are also those of $c$ and of $\alpha_n$ (see Figure 3). Thus $\alpha_n\alpha^{-1}$ is the boundary of an open set contained in $\psi([0,1] \times [-\frac{r}{2^n}, \frac{r}{2^n}])$, and by Proposition 1.29, we have

$$\begin{aligned} d_P(H_c, H_\alpha) &\leq d_P(H_c, H_{\alpha_n}) + d_P(H_{\alpha_n}, H_\alpha) \\ &\leq \frac{\varepsilon}{2} + C\sigma\left(\psi\left([0,1] \times \left[-\frac{r}{2^n}, \frac{r}{2^n}\right]\right)\right)^{\frac{1}{2}} \\ &\leq \varepsilon. \end{aligned}$$

□

PROOF OF PROPOSITION 2.35. Let $\alpha$ be a piecewise geodesic path with the same endpoints as $c$ and let us fix a parametrization of $\alpha$.

Let $\delta_0$ be the distance between $c(0)$ and $c(1)$. Assume that $d_\infty(c,\alpha)$ is smaller than $\inf(r, r'/4, \delta_0/10, R_M/2)$, where $r$ is the radius of the tube around $c$, $r'$ the number given by Lemma 2.28 and $R_M$ the convexity radius of $M$ (see Theorem 2.6).

Consider the two balls $B_0 = B(c(0), 2d_\infty(c,\alpha))$ and $B_1 = B(c(1), 2d_\infty(\alpha,c))$. By definition of $\delta_0$, these balls are disjoint, and, since $\alpha(0) = c(0)$ and $\alpha(1) = c(1)$, there exists a last time $\tau_0$ at which $\alpha$ exits $B_0$ and a first time $\tau_1$ at which it enters $B_1$.

We claim that the points $\alpha(\tau_0)$ and $\alpha(\tau_1)$ are necessarily in the interior of the tube. For one thing, these points are certainly at a distance smaller than $d_\infty(c,\alpha)$ from $c$. Consider $\alpha(\tau_0)$ for example and let $p$ be a point of the range of $c$ which realizes the minimum distance from $m$ to $c$. If $p$ were an endpoint of $c$, we would have $\alpha(\tau_0) \in B(c(0), d_\infty(c,\alpha))$ in contradiction with the definition of $\tau_0$. Thus, $p$ is an interior point of $c$ and this implies that the minimizing geodesic from $\alpha(\tau_0)$ to $p$ is orthogonal to $c$. Hence, if $p = c(t_0)$, $\alpha_0$ is a point of the form $\psi(t_0, s_0)$ for some $s_0$ such that $|s_0| < d_\infty(c,\alpha) < r$ and it belongs to the interior of the tube.

Now we claim that $\tau_0 < \tau_1$. For $d(c(0), c(\tau_0)) \leq d(c(0), \alpha(\tau_0)) + d(\alpha(\tau_0), c(\tau_0)) \leq 3d_\infty(c,\alpha) < r'$, so that, by Lemma 2.28, $c([0, \tau_0])$ is contained in $B(c(0), 3d_\infty(c,\alpha))$. Similarly, $c([\tau_1, 1])$ is contained in $B(c(1), 3d_\infty(c,\alpha))$ and these balls are disjoint, so that $\tau_0 < \tau_1$.

Let $\gamma_0$ be the minimizing geodesic from $c(0)$ to $\alpha(\tau_0)$ and $\gamma_1$ the minimizing geodesic from $\alpha(\tau_1)$ to $c(1)$. Since $d_\infty(c,\alpha) < R_M/2$, then these geodesics are well defined and are contained in the balls $B_0$ and $B_1$ respectively.

Let us decompose $\alpha$ in the following way:
$$\alpha \simeq \alpha_{|[0,\tau_0]}\gamma_0^{-1} \cdot \gamma_0 \alpha_{|[\tau_0,\tau_1]}\gamma_1 \cdot \gamma_1^{-1}\alpha_{|[\tau_1,1]}.$$

The first and the third terms are small loops that we shall study later. Let us focus on the central term $\gamma_0 \alpha_{|[\tau_0,\tau_1]}\gamma_1$. According to Proposition 2.14, we can write it, up to equivalence, as a product $\lambda_1 \ldots \lambda_p \xi$, where each $\lambda_i$ is a lasso based at $c(0)$, and $\xi$ an injective path between $c(0)$ and $c(1)$.

Note that the path $\xi$ satisfies the hypotheses of Lemma 2.36, and that $d_\infty(c,\xi) < 2d_\infty(c,\alpha)$. Let us fix a positive $\varepsilon$. Lemma 2.36 provides us with a $\delta_1$ such that, if $d_\infty(c,\alpha) < \delta_1/2$, then $d_P(H_c, H_\xi) < \varepsilon/2$. It remains to estimate $d_P(H_\xi, H_\alpha)$.

By the invariance properties of the distance on $G$, we have

$$\begin{aligned} d_P(H_\xi, H_\alpha) &= d_P(H_\xi, H_{\alpha_{|[0,\tau_0]}\gamma_0^{-1}} H_{\lambda_1} \ldots H_{\lambda_p} H_\xi H_{\gamma_1^{-1}\alpha_{|[\tau_1,1]}}) \\ &\leq \sum_{i=1}^p d_P(1, H_{\lambda_i}) + d_P(1, H_{\alpha_{|[0,\tau_0]}\gamma_0^{-1}}) + d_P(1, H_{\gamma_1^{-1}\alpha_{|[\tau_1,1]}}). \end{aligned}$$

The paths $\alpha_{|[0,\tau_0]}\gamma_0^{-1}$ and $\gamma_1^{-1}\alpha_{|[\tau_1,1]}$ are loops with finite self-intersection, so, by Proposition 2.16, it is sufficient to control their length. According to Proposition 2.16 again, we also need to estimate the lengths of the buckles of the lassos $\lambda_1, \ldots, \lambda_p$.

We know already by Lemma 2.22 that we may assume $\ell(\xi) \geq \ell(c) - \varepsilon/8$, provided $\delta_1$, and hence $d_\infty(c,\xi)$, are small enough. If we impose now $d_\ell(c,\alpha) < \delta_1$, instead of $d_\infty(c,\alpha) < \delta_1$, then we also get $\ell(\alpha) < \ell(c) + \varepsilon/8$.

Then $0 < \ell(\alpha) - \ell(\xi) < \varepsilon/4$. Let us denote by $\beta_1, \ldots, \beta_p$ the buckles of the lassos $\lambda_1, \ldots, \lambda_p$. By Proposition 2.14,

$$\ell(\xi) + \sum_i \ell(\beta_i) \leq \ell(\gamma_0) + \ell(\alpha_{|[\tau_0,\tau_1]}) + \ell(\gamma_1),$$

and so our assumption on $d_\ell(c,\alpha)$ implies

$$\sum_i \ell(\beta_i) + \ell(\alpha_{|[0,\tau_0]}\gamma_0^{-1}) + \ell(\gamma_1^{-1}\alpha_{|[\tau_1,1]}) \leq \frac{\varepsilon}{4} + 2\ell(\gamma_0) + 2\ell(\gamma_1).$$

Since $\ell(\gamma_i) \leq 2d_\infty(c,\alpha)$, the lengths appearing in the right hand side can be made arbitrarily small by taking $d_\infty(c,\alpha)$ small enough. This means exactly that there exists a $\delta_2$ such that $d_\ell(c,\alpha) < \delta_2$ implies $d_P(H_\xi, H_\alpha) < \varepsilon/2$, and this finishes the proof. □

COROLLARY 2.37. *Let $(\beta_n)_{n\geq 0}$ be any sequence of piecewise geodesic paths with the same endpoints as $c$ such that $d_\ell(\beta_n, c) \longrightarrow 0$. Then the sequence $(H_{\beta_n})_{n\geq 0}$ converges to $H_c$ in the topology induced by $d_P$.*

By Lemma 2.32, there exist such sequences of piecewise geodesic paths. This proves that the variable $H_c$ does not depend on the particular choice of the sequence of paths approximating $c$. Nevertheless, it may still depend on our choice of a Riemannian metric on $M$.

**2.6.3. Continuity of the double layer potential (1).** According to the remark made at the end of Paragraph 2.4.3, Lemma 2.36 and Proposition 2.35 give rise to similar statement for the double layer potential. We can even drop the assumption that $\alpha$ be piecewise geodesic in the lemma.

LEMMA 2.38. *For all $\varepsilon > 0$, there exists $\delta > 0$ such that if $\alpha$ is an injective path with the same endpoints as $c$, such that $d_\infty(c, \alpha) < \delta$ and $\alpha((0, 1)) \subset \psi((0, 1) \times (-r, r))$, then $\| u_\alpha - u_c \|_{L^2} < \varepsilon$.*

PROPOSITION 2.39. *For all $\varepsilon > 0$, there exists $\delta > 0$ such that if $\alpha$ is a piecewise geodesic path with the same endpoints as $c$ and such that $d_\ell(c, \alpha) < \delta$, then $\| u_\alpha - u_c \|_{L^2} < \varepsilon$.*

### 2.6.4. Random holonomy along arbitrary paths.

We have now at our disposal one random variable on the probability space $(\Omega, \mathcal{A}, P_x)$ for each piecewise geodesic path and one for each embedded path. The paths which we want to consider are concatenations of a finite number of embedded paths, so we can certainly make the following attempt to extend our family.

Let $c \in PM$ be a path. There exist $p \geq 1$ and embedded paths $c_1, \ldots, c_p$ such that $c = c_1 \ldots c_p$. The random variable $H_{c_p} \ldots H_{c_1}$ is well defined, but depends a priori on the decomposition of $c$, which is far from unique.

The following proposition will imply the expected independence with respect to the decomposition as an immediate corollary.

PROPOSITION 2.40. *Let $(\alpha_n)$ be a sequence of piecewise geodesic paths converging with fixed endpoints to $c$. The sequence $(H_{\alpha_n})$ converges to the product $H_{c_p} \ldots H_{c_1}$.*

COROLLARY 2.41. *The product $H_{c_p} \ldots H_{c_1}$ is independent of the choice of the decomposition of $c$ and it is equal to the common limit of all sequences $(H_{\alpha_n})$ associated with sequences $(\alpha_n)$ of piecewise geodesic paths converging to $c$ with fixed endpoints. We shall denote it by $H_c$.*

Note that there exist such sequences as $(\alpha_n)$ by Lemma 2.32.

PROOF OF PROPOSITION 2.40. Let $(\alpha_n)$ be a sequence of piecewise geodesic paths converging to $c$. We want to cut each $\alpha_n$ in a way which corresonds to the decomposition of $c$. Let us fix a parametrization of $c$ such that $c_i = c_{|[\frac{i-1}{p}, \frac{i}{p}]}$ for $i = 1, \ldots, p$. Let us also fix a parametrization of each $\alpha_n$ such that the uniform convergence $\sup_t(c(t), \alpha_n(t)) \to 0$ holds. Finally, set $\alpha_{i,n} = \alpha_{n|[\frac{i-1}{p}, \frac{i}{p}]}$. Let us show that for each $i = 1, \ldots, p$, $d_\ell(\alpha_{i,n}, c_i) \xrightarrow[n \to \infty]{} 0$.

The first point is that $\sup_t(c_i(t), \alpha_{i,n}(t)) \leq \sup_t(c(t), \alpha_n(t)) \xrightarrow[n \to \infty]{} 0$, so that $d_\infty(\alpha_{i,n}, c_i) \longrightarrow 0$.

That $\ell(\alpha_{i,n}) \to \ell(c_i)$ for all $i$ follows from the last statement of Lemma 2.22.

Consider now for each $i$ and each $n$ the path $\tilde{\alpha}_{i,n}$ which is $\alpha_{i,n}$ concatenated at both ends with a minimizing piece of geodesic in order to have the same end points as $c_i$. If $n$ is large enough, $\alpha_n$ is close enough to $c$ for these pieces of minimizing geodesic to be uniquely defined and these geodesic segments cancel out in the product $\tilde{\alpha}_{1,n} \ldots \tilde{\alpha}_{p,n}$, which is therefore equivalent to $\alpha_{1,n} \ldots \alpha_{p,n}$. On the other hand, the small geodesic pieces stay close to each $c_i$ and their length converges to zero so that $\tilde{\alpha}_{i,n} \xrightarrow{\ell} c_i$ for all $i$. Thus, Corollary 2.37 implies $H_{\tilde{\alpha}_{i,n}} \longrightarrow H_{c_i}$, and

$$H_{\alpha_n} = H_{\tilde{\alpha}_{p,n}} \ldots H_{\tilde{\alpha}_{1,n}} \xrightarrow[n \to \infty]{} H_{c_p} \ldots H_{c_1},$$

which proves the result. □

### 2.6.5. A continuity result and a first summary.

The continuity result that we are about to state and prove is more that a mere extension of Proposition 2.35, because it does not involve piecewise geodesic paths anymore. It is an important step towards the proof that the distribution of the family of variables $(H_c)_{c \in PM}$ that we have now at hand does not depends on the choice of the Riemannian metric.

PROPOSITION 2.42. *Let $c$ be a path of $PM$. For any $\varepsilon > 0$, there exists $\delta > 0$ such that if $c'$ is another path of $PM$ with the same endpoints as $c$, and if $d_\ell(c, c') < \delta$, then $d_P(H_c, H_{c'}) < \varepsilon$.*

PROOF. Let us fix $\varepsilon > 0$ and consider $\delta_0 > 0$ given by Proposition 2.35 such that for any piecewise geodesic path $\alpha$ with the same endpoints as $c$, $d_\ell(c, \alpha) < \delta_0$ implies $d_P(H_c, H_\alpha) < \varepsilon/2$. We claim that Proposition 2.42 holds with $\delta = \delta_0/2$.

Suppose that $c'$ is a path of $PM$ with the same endpoints as $c$ and such that $d_\ell(c, c') < \delta$. We can approach $c'$ by piecewise geodesic paths as closely as we want and this with convergence of the random holonomies. So there exists $\alpha$ a piecewise geodesic path such that, at once, $d_\ell(\alpha, c') < \delta$ and $d_P(H_{c'}, H_\alpha) < \frac{\varepsilon}{2}$. Then this implies $d_\ell(c, \alpha) < \delta_0$, so that

$$d_P(H_c, H_{c'}) \leq d_P(H_c, H_\alpha) + d_P(H_\alpha, H_{c'}) < \varepsilon. \qquad \square$$

Let us summarize the results of the procedure of piecewise geodesic approximation. We have indeed reached the center of the continuum limit procedure and what remains to do now is essentially to check that the construction does not depend on the choice of the metric and that it is consistent with the discrete theory.

We are now putting together Propositions 2.40, 2.41 and 2.42. Consider the probability space $(\Omega, \mathcal{A}, P_x)$ defined in Definition 2.8.

PROPOSITION 2.43. *Let $c$ be a path of $PM$.*

*1. Let $(\alpha_n)_{n \geq 0}$ be a sequence of piecewise geodesic paths converging in length to $c$ with fixed endpoints. Then the sequence $(H_{\alpha_n})_{n \geq 0}$ converges in probability to a random variable which depends only on $c$ and which we denote by $H_c$.*

*2. Moreover, for any $\varepsilon > 0$, there exists $\delta > 0$ such that if $c'$ is another path of $PM$ with the same endpoints as $c$ and if $d_\ell(c, c') < \delta$, then $d_P(H_c, H_{c'}) < \varepsilon$.*

*3. The random variable associated to the path $c^{-1}$ satisfies $H_{c^{-1}} = H_c^{-1}$ almost surely.*

*4. If $c_1$ and $c_2$ satisfy $c_1(1) = c_2(0)$, then $H_{c_1 c_2} = H_{c_2} H_{c_1}$ almost surely.*

*5. If $c$ and $c'$ are two equivalent paths in the sense of Definition 2.12, then $H_c = H_{c'}$ almost surely.*

PROOF. Statements 1 and 2 have already been proved. By reversing the orientation of a sequence of piecewise geodesic paths approximating $c$, we see that 3 holds. Property 4 is proved by concatenating sequences approximating of $c_1$ and $c_2$. Finally, 5 follows from 3 and 4. $\qquad \square$

### 2.6.6. Continuity of the double layer potential (2).

Once again, we can transpose directly previous arguments to the double layer potential and get the following result.

PROPOSITION 2.44. *Let $c$ be a path of $PM$. For all $\varepsilon > 0$, there exists $\delta > 0$ such that if $c'$ is another path of $PM$ with the same endpoints as $c$ and if $d_\ell(c, c') < \delta$, then $\| u_c - u_{c'} \|_{L^2} < \varepsilon$.*

In this particular context, and if we consider loops, we can drop the assumption on the endpoints. We denote by $LM$ the subset of $PM$ consisting of loops.

COROLLARY 2.45. *Let $l$ be a loop of $LM$. For all $\varepsilon > 0$, there exists $\delta > 0$ such that if $l'$ is another loop of $LM$ and if $d_\ell(l, l') < \delta$, then $\| u_l - u_{l'} \|_{L^2} < \varepsilon$.*

PROOF. Let $\delta_0$ be given by the preceding proposition and set $\delta = \frac{\delta_0}{3}$. Let $l' \in LM$ be such that $d_\ell(l, l') < \delta$. We have in particular $d(l(0), l'(0)) < \delta$. Let now $\sigma$ be a minimizing geodesic from $l(0)$ to $l'(0)$. Then $\tilde{l}' = \sigma l' \sigma^{-1}$ satisfies $u_{\tilde{l}'} = u_{l'}$ a.e. and $d_\ell(\tilde{l}', l) < \delta_0$. Moreover, $\tilde{l}'$ has the same endpoints as $l$. Thus,

$$\| u_l - u_{l'} \|_{L^2} = \| u_l - u_{\tilde{l}'} \|_{L^2} < \varepsilon. \qquad \square$$

## 2.7. Consistency with the discrete theory

Given a graph on $M$ and a finite collection $c_1, \ldots, c_n$ of paths in this graph, we expect the distribution of $(H_{c_1}, \ldots, H_{c_n})$ under $P_{(x_1, \ldots, x_q)}$, which is now a well-defined probability measure on $G^n$, to be the same as the distribution of $(h_{c_1}, \ldots, h_{c_n})$ under $P^\Gamma_{(x_1, \ldots, x_q)}$. For the moment, we know only that this is true when the paths are piecewise geodesic and we need to lift this restriction. This goes actually hand in hand with the proof that the distribution of the family $(H_c)_{c \in PM}$ does not depend on the choice of the Riemannian metric. The key result is the following. Recall that $E_x$ denotes the expectation with respect to $P_{(x_1, \ldots, x_q)}$.

PROPOSITION 2.46. *Let $\Gamma = \{e_1, \ldots, e_r\}$ be a graph on $M$ such that the loops $L_1, \ldots L_q$ belong to $\Gamma^*$. For any function $f$ continuous on $G^\Gamma$, we have:*

$$E_x f(H_{e_1}, \ldots, H_{e_r}) = \int_{G^\Gamma} f \, dP^\Gamma_{(x_1, \ldots, x_q)}.$$

COROLLARY 2.47. *Let $c_1, \ldots, c_n$ be a family of paths. Assume that there exists a graph $\Gamma$ such that $c_1, \ldots, c_n$ and $L_1, \ldots, L_q$ belong to $\Gamma^*$. Then, for any function $f$ continuous on $G^n$, we have:*

$$E_x f(H_{c_1}, \ldots, H_{c_n}) = \int_{G^\Gamma} f(h_{c_1}, \ldots, h_{c_n}) \, dP^\Gamma_{(x_1, \ldots, x_q)}.$$

PROOF. This follows immediately from the previous proposition by using the multiplicativity of the random holonomy. $\qquad \square$

As a consequence, we get the following independence result.

COROLLARY 2.48. *Consider two Riemannian metrics on $M$ consistent with the surface measure $\sigma$ and such that $L_1, \ldots, L_q$ are geodesics. Each of these metrics allows us to define a family of $G$-valued random variables indexed by $PM$, $(H_c^1)_{c \in PM}$ and $(H_c^2)_{c \in PM}$. Then these two families of random variables have the same distribution, in that, for any finite family $c_1, \ldots, c_n$ of paths on $M$, the distributions of $(H_{c_1}^1, \ldots, H_{c_n}^1)$ and $(H_{c_1}^2, \ldots, H_{c_n}^2)$ are equal as probability measures on $G^n$.*

PROOF. Let $\zeta_1, \ldots, \zeta_n$ be a family of paths which are piecewise geodesic for the first metric. There exists a graph such that $\zeta_1, \ldots, \zeta_n$ and $L_1, \ldots, L_q$ belong to $\Gamma^*$, so, by Corollary 2.47, $(H_{\zeta_1}^1, \ldots, H_{\zeta_n}^1)$ and $(H_{\zeta_1}^2, \ldots, H_{\zeta_n}^2)$ have the same distribution, because they both have that of $(h_{\zeta_1}, \ldots, h_{\zeta_2})$ under $P_x^\Gamma$.

Now, by Lemma 2.32, the space of piecewise geodesic paths for the first metric is dense in $PM$. Any family $(c_1,\ldots,c_n)$ is the componentwise limit in length of a sequence of families $(\zeta_{1,k},\ldots,\zeta_{n,k})$ of piecewise geodesic paths for the first metric. By the continuity property 1 of Proposition 2.43, $H^1_{\zeta_{i,k}}$ tends to $H^1_{c_i}$ and $H^2_{\zeta_{i,k}}$ tends to $H^2_{c_i}$ as $k$ tends to infinity, for all $i$. These convergence hold in probability. This implies that the $n$-tuples $(H^1_{\zeta_{1,k}},\ldots,H^1_{\zeta_{n,k}})$ and $(H^2_{\zeta_{1,k}},\ldots,H^2_{\zeta_{n,k}})$ also converge in probability, or in distance $d_P$, as $G^n$-valued variables, respectively to $(H^1_{c_1},\ldots,H^1_{c_n})$ and $(H^2_{c_1},\ldots,H^2_{c_n})$. This implies in particular the convergence of the distributions of the $n$-tuples, hence the result. □

The proof of Proposition 2.46 relies upon the approximation of graphs by piecewise geodesic graphs. Before stating a result, let us gather some elementary facts about the edges and faces of a graph on $M$.

Recall that a path and hence an edge must, by definition, have a non-zero derivative at their endpoints, this to avoid pathological behaviour. For example, in a given graph, consider all edges at a given vertex and a small geodesic circle centered at this vertex. If the radius of this circle is small enough, then, by Lemma 2.28, each edge cuts it only once, transversally, and the ordering of the intersection points around the circle, which does not depend on the radius of the circle, defines a cyclic order on the set of these edges.

Now, consider two edges which are adjacent for this order. They bound at least one common face. Thus, if $M$ is oriented, a couple of adjacent edges determines a face of the graph. Conversely, given a face, any two consecutive edges of the boundary of this face are adjacent at the vertex that they share, or maybe at both vertices if they share two.

Let us choose an orientation of $M$. Recall that we still have a Riemannian metric on $M$.

PROPOSITION 2.49. *Let $\Gamma = \{e_1,\ldots,e_r\}$ be a graph such that $L_1,\ldots,L_q \in \Gamma^*$. For any $\varepsilon > 0$, there exists a graph $\Gamma_\varepsilon = \{\eta_1,\ldots,\eta_r\}$ with piecewise geodesic edges, such that $L_1,\ldots,L_q \in \Gamma^*_\varepsilon$, and such that:*

*1. $\Gamma_\varepsilon$ and $\Gamma$ have the same vertices,*

*2. for each $i = 1,\ldots,r$, $\eta_i$ has the same endpoints as $e_i$ and $d_\ell(\eta_i,e_i) < \varepsilon$,*

*3. for each $i = 1,\ldots,r$, if the edge $e_i$ is involved in the decomposition of some $L_j$, then $\eta_i = e_i$,*

*4. for each $i = 1,\ldots,r$, if the edge $e_i$ is not involved in the decomposition of any $L_j$, then the interior of $\eta_i$ is fully contained in the same connected component of the complement of the ranges of $L_1,\ldots,L_q$ which contains the interior of $e_i$.*

*Let us denote by $\eta : \Gamma^* \longrightarrow \Gamma^*_\varepsilon$ the multiplicative map which sends $e_i$ to $\eta_i$. It is possible to construct $\Gamma_\varepsilon$ in such a way that this map induces a one-to-one correspondence, still denoted by $\eta : \mathcal{F}(\Gamma) \longrightarrow \mathcal{F}(\Gamma_\varepsilon)$, between the faces of $\Gamma$ and those of $\Gamma_\varepsilon$, such that $\partial \eta(F) = \eta(\partial F)$ and $\sigma(F - \eta(F)) < \varepsilon$, where $-$ denotes the symmetric difference.*

PROOF. Although property 4 is the longest to state, it is a simple consequence of 1 and 2. Indeed, if the interior of $e_i$ is in a given connected component of $M - \bigcup_j L_j([0,1])$, then by property 2, $\eta_i$ meets this component, provided $\varepsilon$ is small enough. If $\varepsilon$ is small enough again, properties 1 and 2 imply that $\eta_i$ and $e_i$ have the same endpoints. Since a crossing between $\eta_i$ and a loop $L_j$ could happen only at a vertex of the graph, and since, by 1, the only vertices on $\eta_i$ are its endpoints, the interior of $\eta_i$ is contained in the same component as $e_i$.

Let $\mathcal{V}(\Gamma) = \{v_1, \ldots, v_p\}$ denote the set of vertices of $\Gamma$. Fix $\varepsilon > 0$ and let $r$ be a positive real number which 'localizes' the vertices of $\Gamma$ in the sense that the following properties hold.

(i) The balls $B(v_i, r)$ are pairwise disjoint.

(ii) For every pair $(e_i, v_j)$ consisting of an edge $e_i$ and an endpoint $v_j$ of $e_i$, $e_i$ meets only once and transversally any circle centered at $v_j$ and of radius smaller than $r$. Moreover, the length of the portion of $e_i$ in the corresponding ball is smaller than $\varepsilon/16$.

(iii) For any pair $(e_i, v_j)$ where $v_j$ is not an endpoint of $e_i$, $e_i$ does not meet the ball $B(v_j, r)$.

(iv) The sum of the areas of the balls $B(v_i, r)$ is smaller than $\varepsilon/2$.

(v) $r < \varepsilon/16$ and $r < R_M$, where $R_M$ is the convexity radius of $M$ defined in Theorem 2.6.

Each of these properties holds for $r$ small enough and remains true when $r$ is replaced by $r' < r$ once it holds for some $r$, so that it is not a problem to get all of them to hold simultaneously.

Let $t$ be a positive real number such that the area $\sigma(\{d(\cdot, \Gamma) < t\})$ of the set of points at distance smaller than $t$ from $\Gamma$ is smaller than $\varepsilon/2$. For each $i$, let $\widehat{e}_i$ denote the portion of $e_i$ outside the two disks of radius $r$ around its endpoints. Let $\delta$ be the smallest distance between the images of two distinct paths $\widehat{e}_i$. For each $i$, let $\gamma_i$ be an injective piecewise geodesic path with the same endpoints as $\widehat{e}_i$, such that $d_\ell(\gamma_i, \widehat{e}_i) < \inf(\varepsilon/4, \delta/2, t)$, and which never meets the balls $B(v_j, r)$, except at its ends. This last condition can be obtained because $e_i$ cuts $B(v_j, r)$ transversally if $v_j$ is an endpoint of $e_i$, so that, in a neighbourhood of each end point of $\widehat{e}_i$, there is a half-tube around $\widehat{e}_i$ which does not meet $B(v_j, r)$ and in which one can perform the approximation of $\widehat{e}_i$. Note that, by definition of $\delta$, the $\gamma_i$'s are disjoint.

Now define $\eta_i$ for each $i$ such that $e_i$ is not already piecewise geodesic as the concatenation of the minimizing geodesic from $e_i(0)$ to $\widehat{e}_i(0)$, of $\gamma_i$ and of the minimizing geodesic from $\widehat{e}_i(1)$ to $e_i(1)$ (see fig. 4). Assumption (v) on $r$ ensures that these minimizing geodesics are well defined. Assumptions (ii) and (v) imply that $d_\ell(\eta_i, e_i) < \varepsilon$.

As for the edges of the decompositions of $L_1, \ldots, L_q$, they are already piecewise geodesic and it suffices to rename them, setting $\eta_i = e_i$.

The $\eta_i$'s are injective piecewise geodesic paths, so they are edges. Moreover, they have been constructed in such a way that they meet each other only at their endpoints: we have already noticed that they could not meet outside the balls around the vertices of $\Gamma$, and they cannot either meet more than once inside these balls, according to the local properties of geodesics. Thus, the graph $\Gamma_\varepsilon = \{\eta_1, \ldots, \eta_r\}$ exists and has the same vertices as $\Gamma$.

Properties 1, 2 and 3 are true by construction of $\Gamma_\varepsilon$, and they imply property 4 as explained at the beginning of the proof. It remains to prove the last part of the statement.

Consider the set of edges of $\Gamma$ at a given vertex and the cyclic order on this set induced by the orientation of $M$. By construction, the corresponding edges of $\Gamma_\varepsilon$ cut the circle of radius $r$ around this vertex in the same cyclic order, so that the

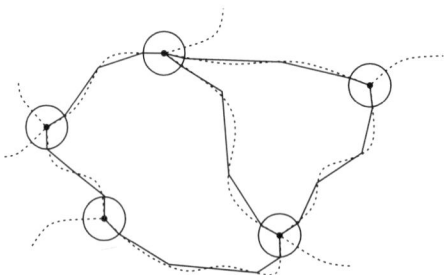

FIGURE 4. Definition of the edges of the graph $\Gamma_\varepsilon$.

multiplicative application $\eta : \Gamma^* \to \Gamma_\varepsilon^*$ defined by $\eta(e_i) = \eta_i$ preserves the cyclic order at each vertex.

Given a pair of edges of $\Gamma$ which determine the face $F$, the pair of corresponding edges of $\Gamma_\varepsilon$ is a pair of adjacent edges hence it determines a face of $\Gamma_\varepsilon$. It is easily checked that this face does not depend on the particular choice of the pair of edges that represents $F$ and we denote it by $\eta(F)$. By construction again, we have the relation $\partial(\eta(F)) = \eta(\partial F)$.

The symmetric difference of $F$ and $\eta(F)$ is contained in the union of the balls $B(v_j, r)$ and the sets $\{d(\cdot, \hat{e}_i) < t\}$. By assumption (iv) and by definition of $t$, we know that the total volume of these sets is smaller than $\varepsilon$. Thus, $\sigma(F - \eta F) < \varepsilon$. Moreover, if $\varepsilon < \sigma(M)/2$, this inequality characterizes $\eta(F)$ among the faces of $\Gamma_\varepsilon$ which have $\eta(\partial F)$ as boundary, if there is more than one. □

PROOF OF PROPOSITION 2.46. Let $\Gamma$ be a graph on $M$ such that $L_1, \ldots, L_q$ belong to $\Gamma^*$. For each integer $n$, the preceding proposition provides us with a graph $\Gamma_n = \{\eta_{1,n}, \ldots, \eta_{r,n}\}$ corresponding to $\varepsilon = 1/n$. For each $i = 1, \ldots, r$, the sequence $(\eta_{i,n})$ converges in length to $e_i$ with fixed endpoints, so that $H_{\eta_{i,n}} \longrightarrow H_{e_i}$ in distance $d_P$. As explained in the proof of Corollary 2.48, this implies the following convergence in distribution:

$$(H_{\eta_{1,n}}, \ldots, H_{\eta_{r,n}}) \xrightarrow[n \to \infty]{(d)} (H_{e_1}, \ldots, H_{e_r}).$$

Thus, for any function $f$ continuous on $G^\Gamma$,

$$E_x f(H_{e_1}, \ldots, H_{e_r}) = \lim_{n \to \infty} \frac{1}{Z^{\Gamma_n}(x)} \int_{G^\Gamma} f(g_1, \ldots, g_r) \prod_{F \in \mathcal{F}(\Gamma_n)} p_{\sigma(F)}(h_{\partial F})$$
$$d\nu_{x_1} \ldots d\nu_{x_q} \, dg'$$

$$= \lim_{n \to \infty} \frac{1}{Z^{\Gamma_n}(x)} \int_{G^{\Gamma_n}} f(g_1, \ldots, g_r) \prod_{F \in \mathcal{F}(\Gamma)} p_{\sigma(\eta_n(F))}(h_{\partial \eta_n(F)})$$
$$d\nu_{x_1} \ldots d\nu_{x_q} \, dg',$$

$$= \left(\lim_{n \to \infty} \frac{1}{Z^{\Gamma_n}(x)}\right) \int_{G^\Gamma} f(g_1, \ldots, g_r) \prod_{F \in \mathcal{F}(\Gamma)} p_{\sigma(F)}(h_{\partial F})$$
$$d\nu_{x_1} \ldots d\nu_{x_q} \, dg',$$

because, for each face $F$ of $\Gamma$, $\sigma(\eta_n(F))$ tends to $\sigma(F)$ when $n$ tends to infinity. We have used $Z(x)$ as a shorthand notation for $Z(x_1,\ldots,x_q)$.

By Proposition 1.26, the conditional partition functions computed in two graphs, one finer than the other, are equal. But given two piecewise geodesic graphs, there exists a third one which is finer than both others, according to Proposition 2.5. Thus the partition function is the same for all piecewise geodesic graphs, and $Z^{\Gamma_n}$ is independent of $n$. Taking $f$ constant, we see that it is equal to $Z^\Gamma$. □

The following result about the partition functions, which has been proved incidentally, is very important.

PROPOSITION 2.50. *Let $\Gamma$ be a graph such that $L_1,\ldots,L_q \in \Gamma^*$. Then the value of the conditional partition function $Z^\Gamma(x_1,\ldots,x_q)$ does not depend on $\Gamma$.*

We have not used yet the properties 3 and 4 of the piecewise geodesic approximation of graphs. They are actually designed to allow us to extend the construction of the random holonomy to surfaces with boundary.

## 2.8. Binding up the conditional versions

What we have done so far is, for each choice of $x = (x_1,\ldots,x_q) \in G^q$, to construct a family of random variables $(H_c)_{c \in PM}$ on the probability space $(\Omega, \mathcal{A}, P_{(x_1,\ldots,x_q)})$. But the random variable associated to a path is defined only $P_{(x_1,\ldots,x_q)}$-almost surely, and needs not be the same for two different choices of $x$.

This is especially annoying because we expect a result similar to Corollary 1.15 to hold, saying that $(x_1,\ldots,x_q) \longmapsto P_{(x_1,\ldots,x_q)}$ provides a disintegration of $P$ with respect to $(H_{L_1},\ldots,H_{L_q})$, and also to be able to apply this result when we consider variables $H_c$ for non-piecewise geodesic paths $c$. This requires to find a single probability space on which all measures $P_x$ and all variables $H_c$ are defined at the same time.

The canonical measurable space on which all variables $H_c$ are simultaneously defined is the set $\mathcal{F}(PM,G)$ of all maps from $PM$ to $G$ endowed with the cylinder $\sigma$-field $\mathcal{C}$. For each disintegration family $L_1,\ldots,L_q$ and each $x = (x_1,\ldots,x_q) \in G^q$, the distribution of $(H_c)_{c \in PM}$ under $P_{(x_1,\ldots,x_q)}$ is a probability measure on $(\mathcal{F}(PM,G), \mathcal{C})$ which we shall denote by $P^c_{(x_1,\ldots,x_q)}$, where $c$ stands for *canonical*.

The price for putting the measures together is that we define them on the least possible informative probability space, as far as the properties of the sample paths $c \mapsto H_c(\omega)$ are concerned. We are merely facing the fact that we have defined no more than the finite dimensional marginals of a process.

A good point is that we can prove the disintegration property. Let us state separately the following consequence of Lemma 2.32, in which $M$ is allowed to have a boundary.

LEMMA 2.51. *Let $L_1,\ldots,L_q$ be a disintegration family and $(c_1,\ldots,c_n)$ a finite collection of paths. There exists a sequence $(c_{1,k},\ldots,c_{n,k})_{k \geq 0}$ of collections of paths such that*

*1. for each $k$, there exists a graph $\Gamma$ such that $L_1,\ldots,L_q$ and $c_{1,k},\ldots,c_{n,k}$ belong to $\Gamma^*$,*

*2. for each $i = 1,\ldots,n$, $c_{i,k} \xrightarrow{\ell} c_i$ with fixed endpoints.*

PROPOSITION 2.52. *Let $L_1, \ldots, L_q$ be a disintegration family on $M$. The map $(x_1, \ldots, x_q) \longmapsto P^c_{(x_1, \ldots, x_q)}$ is a disintegration of the measure $P^c$ with respect to the random variable $(H_{L_1}, \ldots, H_{L_q})$ on the measurable space $(\mathcal{F}(PM, G), \mathcal{C})$. This means that*

1. $(H_{L_1}, \ldots, H_{L_q}) = (x_1, \ldots, x_q)$ $P^c_{(x_1, \ldots, x_q)}$-*a.s.*
2. *If $\eta$ denotes the distribution of $(H_{L_1}, \ldots, H_{L_q})$ under $P^c$, we have*

$$P^c = \int_{G^q} P^c_{(x_1, \ldots, x_q)} \, d\eta(x_1, \ldots, x_q).$$

*The following property also holds:*

2'. *If $0 < r \leq q$ and $\eta_r$ denotes the distribution of $(H_{L_1}, \ldots, H_{L_r})$ under $P^c_{(x_{r+1}, \ldots, x_q)}$, then*

$$P^c_{(x_{r+1}, \ldots, x_q)} = \int_{G^r} P^c_{(x_1, \ldots, x_q)} \, d\eta_r(x_1, \ldots, x_r),$$

*where $x_i$ always refers to the loop $L_i$.*

PROOF. By construction, the relation $(H_{L_1}, \ldots, H_{L_q}) = (x_1, \ldots, x_q)$ holds almost surely on the probability space $(\Omega, \mathcal{A}, P_{(x_1, \ldots, x_q)})$ for all $x_1, \ldots, x_q$, and this implies property 1.

Property 2 is a special case of property 2', when $r = q$. Because we are working with the cylindrical $\sigma$-field, we need only prove that, for each $n$, each collection $c_1, \ldots, c_n$ of paths and each continuous function $f$ on $G^n$,

$$E^c_{(x_{r+1}, \ldots, x_q)} f(H_{c_1}, \ldots, H_{c_n}) = \int_{G^q} E^c_{(x_1, \ldots, x_q)} f(H_{c_1}, \ldots, H_{c_n}) \, d\eta(x_1, \ldots, x_r). \tag{2.4}$$

If there exists a graph $\Gamma$ such that $L_1, \ldots, L_q$ and $c_1, \ldots, c_n$ belong to $\Gamma^*$, then the consistency with the discrete theory of each measure $P_{(x_1, \ldots, x_q)}$ (Proposition 2.47) and the disintegration property on a graph (Corollary 1.15) imply the result.

Otherwise, Lemma 2.51 allows us to approximate each path $c_i$ by a sequence $(c_{i,k})$ such that (2.4) holds when each $c_i$ is replaced by $c_{i,k}$. By property 2 in Proposition 2.43, the fact that $d_\ell(c_{i,k}, c_i) \longrightarrow 0$ with fixed endpoints implies the convergence of the left-hand side of (2.4) and the pointwise convergence of the integrand on the right-hand side, and the result is proved. □

## 2.9. Surfaces with boundary

It is time now to extend the construction to the case of a surface with boundary. Let us consider such a surface $M$. We are going to construct a random holonomy on $M$ by embedding it into a minimal closure (see Definition 1.1) and by restricting the random holonomy on this minimal closure to the paths of $M$. The work is essentially already done, but still we need to check that the resulting random holonomy does not depend on the choice of the minimal closure of $M$. This is actually the first instance of the Markov property of the random holonomy, which will be studied more extensively in Chapter 5.

It seems to be the easiest route to construct first the conditional random holonomy with deterministic boundary conditions and then to integrate properly these conditional versions to define a 'free' random holonomy.

**2.9.1. Conditional random holonomy.** Let $M$ be a surface with boundary and let $N_1, \ldots, N_p$ be the connected components of its boundary. Let $L_1, \ldots, L_q$ be a disintegration family on $M$ such that each $L_i$ is contained in the interior of $M$. Consider a minimal closure $i : M \longrightarrow M_1$ of $M$. Recall that $i$ is an embedding of $M$ into a closed surface $M_1$ such that $M_1 - i(M)$ is the union of $p$ open disks. We will identify $M$ with a compact submanifold of $M_1$.

The family $L_1, \ldots, L_q, N_1, \ldots, N_p$ is a disintegration family on $M_1$. Let $x = (x_1, \ldots, x_q)$ and $y = (y_1, \ldots, y_p)$ be families of elements of $G$ and let us consider the measure $P^{M_1}_{(x_1,\ldots,x_q,y_1,\ldots,y_p)}$ on $\mathcal{F}(PM_1, G)$. The superscript $c$ is not necessary anymore, since we are working on the canonical space anyway, but we need to remember which surface we are considering.

Seeing $M$ as a submanifold of $M_1$ allows us also to see $PM$ as a subspace of $PM_1$. If we define the convergence in length on $PM_1$ using a certain metric on $M_1$ and that on $PM$ using the restriction of the same metric, we see that this inclusion preserves the convergence of paths.

Now there is a very simple map from $\mathcal{F}(PM_1, G)$ to $\mathcal{F}(PM, G)$ which is the restriction map and we denote by $P^M_{(x_1,\ldots,x_q;y_1,\ldots,y_q)}$ the push-forward of the measure $P^{M_1}_{(x_1,\ldots,x_q,y_1,\ldots,y_p)}$ by this restriction. Equivalently, $P^M_{x;y}$ is the distribution of the family $(H_c)_{c \in M}$ under the probability measure $P^{M_1}_{x,y}$. The reason for which a semi-colon has appeared in the subscript is that the boundary conditions and the conditions on interior loops do not play the same role on $M$ and we want to distinguish them.

We must prove that the measure $P^M_{x;y}$ does not depend on the choice of $M_1$. This will be a consequence of the following consistency result.

PROPOSITION 2.53. *Let $\Gamma = \{e_1, \ldots, e_r\}$ be a graph on $M$ such that the loops $L_1, \ldots L_q$ belong to $\Gamma^*$. For any function $f$ continuous on $G^\Gamma$, we have:*

$$E_{x;y} f(H_{e_1}, \ldots, H_{e_r}) = \int_{G^\Gamma} f \, dP^\Gamma_{(x_1,\ldots,x_q,y_1,\ldots,y_p)},$$

*where $E_{x;y}$ denotes the expectation with respect to $P^M_{(x_1,\ldots,x_q;y_1,\ldots,y_p)}$.*

COROLLARY 2.54. *The measure $P^M_{(x_1,\ldots,x_q;y_1,\ldots,y_p)}$ on $(\mathcal{F}(PM, G), \mathcal{C})$ does not depend on the choice of the minimal closure $M_1$ which has been used to define it.*

PROOF. By multiplicativity, Proposition 2.53 implies that, given any collection $(c_1, \ldots, c_n)$ of paths for which there exists a graph $\Gamma$ such that $c_1, \ldots, c_n$ and $L_1, \ldots, L_q$ belong to $\Gamma^*$, the distribution of $(H_{c_1}, \ldots, H_{c_n})$ under $P^M_{x;y}$ is the same as that of $(h_{c_1}, \ldots, h_{c_n})$ under $P^\Gamma_{x,y}$. Lemma 2.51 and the continuity property of the random holonomy (see Proposition 2.43) show that this determines uniquely the distribution of $(H_{c_1}, \ldots, H_{c_n})$ for any finite collection $(c_1, \ldots, c_n)$ of paths, and hence the measure $P^M_{x;y}$ itself, since we are working on the cylinder $\sigma$-field. This characterization of $P^M_{x;y}$ is independent of $M_1$ and the result is proved. □

PROOF OF PROPOSITION 2.53. Let $\Gamma$ be a graph and $f$ a continuous function on $G^\Gamma$. The fact that $M_1$ is a *minimal* closure of $M$ implies that $\Gamma$ is also a graph on $M_1$, with $p$ new faces, one for each boundary component of $M$. By definition of $P^M_{x;y}$, we have

$$E_{x;y} f(H_{e_1}, \ldots, H_{e_r}) = E^{M_1}_{x,y} f(H_{e_1}, \ldots, H_{e_r}),$$

where $E_{x,y}^{M_1}$ denotes the expectation on $\mathcal{F}(PM_1, G)$ with respect to $P_{x,y}^{M_1}$. The consistency with the discrete theory for surfaces without boundary (Proposition 2.46) allows us to compute the right-hand side. Notice that we are now working on $M_1$, which is endowed with a surface measure $\sigma_1$ such that $\sigma_1(D) = \sigma(D)$ for any domain $D \subset M \subset M_1$.

$$E_{x,y}^{M_1} f(H_{e_1}, \ldots, H_{e_r}) = \int_{G^\Gamma} f \, dP_{x,y}^\Gamma$$

$$= \frac{1}{Z_{M_1}^\Gamma(x,y)} \int_{G^\Gamma} f \prod_{F \in \mathcal{F}(\Gamma)} p_{\sigma_1(F)}(h_{\partial F}) \, d\nu_{x_1} \ldots d\nu_{x_q} d\nu_{y_1} \ldots d\nu_{y_p} dg'.$$

Let us denote by $F_1, \ldots, F_p$ the faces of $\Gamma$ on $M_1$ which are not faces of $\Gamma$ on $M$, in such a way that $N_i$ is, up to orientation, the boundary of $F_i$ for each $i$. We see that

$$E_{x,y}^{M_1} f(H_{e_1}, \ldots, H_{e_r}) = \frac{\prod_{i=1}^p p_{\sigma_1(F_i)}(y_i)}{Z_{M_1}^\Gamma(x,y)} \times$$

$$\int_{G^\Gamma} f \prod_{\substack{F \in \mathcal{F}(\Gamma) \\ F \subset M}} p_{\sigma(F)}(h_{\partial F}) \, d\nu_{x_1} \ldots d\nu_{x_q} d\nu_{y_1} \ldots d\nu_{y_p} dg'$$

$$= \frac{\prod_{i=1}^p p_{\sigma_1(F_i)}(y_i)}{Z_{M_1}^\Gamma(x,y)} Z_M^\Gamma(x;y) \int_{G^\Gamma} f \, dP_{(x_1,\ldots,x_q,y_1,\ldots,y_p)}^\Gamma,$$

where, in the last integral, we see $\Gamma$ as a graph on $M$ again. Taking $f$ constant, we see that

$$Z_{M_1}^\Gamma(x,y) = Z_M^\Gamma(x;y) \prod_{i=1}^p p_{\sigma_1(F_i)}(y_i)$$

and the result is proved. □

For the third time, an important property of the conditional partition functions has arisen during the proof.

LEMMA 2.55. *Let $M$ be a surface with boundary and $M_1$ a minimal closure of $M$. Then, with the notation of the proof above, the following equality holds:*

$$Z_{M_1}^\Gamma(x,y) = Z_M^\Gamma(x;y) \prod_{i=1}^p p_{\sigma_1(F_i)}(y_i).$$

*In particular, $Z_M^\Gamma(x;y)$ is independent of $\Gamma$.*

**2.9.2. Free random holonomy.** Given a disintegration family $L_1, \ldots, L_q$ which contains no component of $\partial M$, and given $x = (x_1 \ldots, x_q)$, we would like to define a measure $P_{(x_1,\ldots,x_q)}$ on $\mathcal{F}(PM, G)$, such that the holonomy along the boudary of $M$ has its 'natural' distribution under $P_{(x_1,\ldots,x_q)}$.

That such a 'natural' distribution exists has to be proved, but it is not difficult. Let $\Gamma$ be a graph on $M$ such that $L_1, \ldots, L_q$ belong to $\Gamma^*$ and choose $x = (x_1, \ldots, x_q)$. By Proposition 1.14 and by the consistency of continuous and discrete theories (Proposition 2.53), we know that the distribution of $(H_{N_{r+1}}, \ldots, H_{N_p})$

under $P^\Gamma_{(x_1,\ldots,x_q)}$ is given by

$$\frac{Z^\Gamma(x_1,\ldots,x_q;y_{r+1},\ldots,y_p)}{Z^\Gamma(x_1,\ldots,x_q)}\,dy_{r+1}\ldots dy_p. \tag{2.5}$$

Since $Z^\Gamma(x_1,\ldots,x_q;y_1,\ldots,y_p)$ does not depend on $\Gamma$, so does

$$Z^\Gamma(x_1,\ldots,x_q;y_{r+1},\ldots,y_p) = \int_{G^r} Z^\Gamma(x_1,\ldots,x_q;y_1,\ldots,y_p)\,dy_1\ldots dy_r$$

for all $r = 0,\ldots,p$.

We have proved the following result.

PROPOSITION 2.56. *Let $M$ be a surface with boundary and $L_1,\ldots,L_q$ a disintegration family on $M$ containing no component of $\partial M$. Choose $x = (x_1,\ldots,x_q)$ and $y = (y_1,\ldots,y_p)$ two families of elements of $G$.*

*1. The conditional partition function $Z^\Gamma(x_1,\ldots,x_q;y_1,\ldots,y_p)$ does not depend on the graph in which one computes it.*

*2. Consider $0 \leq r < p$. The distribution of the random variable $(h_{N_{r+1}},\ldots,h_{N_p})$ under $P^\Gamma_{(x_1,\ldots,x_q;y_1,\ldots,y_r)}$ does not depend on $\Gamma$ either and it is given by*

$$\frac{Z(x_1,\ldots,x_q;y_1,\ldots,y_p)}{Z(x_1,\ldots,x_q;y_1,\ldots,y_r)}\,dy_{r+1}\ldots dy_p.$$

We can define the measure $P^c_{(x_1,\ldots,x_q)}$ on $\mathcal{F}(PM,G)$ by

$$P^c_{(x_1,\ldots,x_q)} = \int_{G^p} \frac{Z(x_1,\ldots,x_q;y_1,\ldots,y_p)}{Z(x_1,\ldots,x_q)} P^c_{(x_1,\ldots,x_q;y_1,\ldots,y_p)}\,dy_1\ldots dy_p.$$

In particular,

$$P^c = \frac{1}{Z}\int_{G^p} Z(y_1,\ldots,y_p) P^c_{(y_1,\ldots,y_p)}\,dy_1\ldots dy_p.$$

More generally, for each $r \in \{0,\ldots,p-1\}$, we can define the probability measure $P^c_{(x_1,\ldots,x_q;y_1,\ldots,y_r)}$ on $\mathcal{F}(PM,G)$ by

$$P^c_{(x_1,\ldots,x_q;y_1,\ldots,y_r)} = \int_{G^{p-r}} \frac{Z(x_1,\ldots,x_q;y_1,\ldots,y_p)}{Z(x_1,\ldots,x_q;y_1,\ldots,y_r)} P^c_{(x_1,\ldots,x_q;y_1,\ldots,y_p)}\,dy_{r+1}\ldots dy_p.$$

This time, there is no need to prove the disintegration property, it holds by construction.

PROPOSITION 2.57. *Let $L_1,\ldots,L_q$ be a disintegration family on $M$. The map $(x_1,\ldots,x_q;y_1,\ldots,y_p) \longmapsto P^c_{(x_1,\ldots,x_q;y_1,\ldots,y_p)}$ is a disintegration of the measure $P^c$ with respect to the random variable $(H_{L_1},\ldots,H_{L_q},H_{N_1},\ldots,H_{N_p})$ on the measurable space $(\mathcal{F}(PM,G),\mathcal{C})$.*

*More generally, for each $r$ such that $0 \leq r \leq p$, and given $y_{r+1},\ldots,y_p$ in $G$, the map $(x_1,\ldots,x_q;y_1,\ldots,y_r) \longmapsto P^c_{(x_1,\ldots,x_q;y_1,\ldots,y_p)}$ is a disintegration of the measure $P^c_{(y_{r+1},\ldots,y_p)}$ with respect to the random variable $(H_{L_1},\ldots,H_{L_q},H_{N_1},\ldots,H_{N_r})$.*

## 2.10. The random holonomy process

### 2.10.1. The distribution of the process.
We are now in position to characterize the Yang-Mills measure and to state its existence and unicity. This measure is just the *distribution* of a $G$-valued random process indexed by the set $PM$ of paths on $M$. In the following statement, there is no reference to a choice of a

version of this process or in other words to the support of the measure. One step in this direction will be made in the next and last section. There is no reference either to gauge transformations. This gap will also be filled in Section 2.10.2.

I have tried to write this theorem in such a way that it can be read and understood with minimal reference to the body of this chapter.

THEOREM 2.58. *Let $M$ be a compact surface, with or without boundary. Consider the measurable space $(\mathcal{F}(PM,G),\mathcal{C})$ of maps from $PM$ to $G$ endowed with the cylinder $\sigma$-field, on which a function $H_c$ is canonically defined for each path $c \in PM$. There exists a probability measure $P$ on $(\mathcal{F}(PM,G),\mathcal{C})$ such that the following properties hold.*

1. *Consider a finite collection $(c_1,\ldots,c_n)$ of paths. If there exists a graph $\Gamma$ on $M$ such that $c_1,\ldots,c_n$ can be written as concatenations of edges of $\Gamma$, then the distribution of $(H_{c_1},\ldots,H_{c_n})$ under $P$ is equal to the distribution of the discrete random holonomy along $c_1,\ldots,c_n$ (see Section 1.4).*

2. *For any path $c$ and any sequence $(c_n)_{n\geq 0}$ of paths with the same endpoints as $c$ and converging in length to $c$ (see Definition 2.21), there is convergence of $H_{c_n}$ towards $H_c$, in probability with respect to $P$ (see Definition 2.17).*

*These properties characterize completely the measure $P$, which also satisfies the following properties.*

3. *If $c_1$ and $c_2$ are two paths such that $c_1 c_2$ exists, then $H_{c_1 c_2} = H_{c_2} H_{c_1}$ $P$-almost surely. In particular, if two paths $c$ and $c'$ are equivalent (see Definition 2.12), then $H_c = H_{c'}$ $P$-almost surely.*

4. *Let $\psi : M \longrightarrow M$ be a diffeomorphism which preserves the area, in that $\psi_* \sigma = \sigma$. Then $\psi$ induces a permutation of the set of paths $PM$ which preserves the measure $P$. In other words, for any finite collection $(c_1,\ldots,c_n)$ of paths, the random variables $(H_{c_1},\ldots,H_{c_n})$ and $(H_{\psi(c_1)},\ldots,H_{\psi(c_n)})$ have the same distribution under $P$.*

*Let $N_1,\ldots,N_p$ be the components of $\partial M$, where $p = 0$ if $M$ is closed. Let $L_1,\ldots,L_q$ be a disintegration family of $M$ containing no boundary component, that is, a collection of disjoint simple loops contained in the interior of $M$. Note that $q$ can be equal to zero. For each $x = (x_1,\ldots,x_q) \in G^q$ and $y = (y_1,\ldots,y_p) \in G^p$, there exists a probability measure $P_{(x_1,\ldots,x_q;y_1,\ldots,y_p)}$ on $(\mathcal{F}(PM,G),\mathcal{C})$ such that the following properties hold.*

5. *The map $(x_1,\ldots,x_q;y_1,\ldots,y_p) \longmapsto P_{(x_1,\ldots,x_q;y_1,\ldots,y_p)}$ is a disintegration of the measure $P$ with respect to the random variable $(H_{L_1},\ldots,H_{L_q},H_{N_1},\ldots,H_{N_p})$.*

6. *For each $(x_1,\ldots,x_q)$ and $(y_1,\ldots,y_p)$, properties 2. and 3. hold when $P$ is replaced by $P_{(x_1,\ldots,x_q;y_1,\ldots,y_p)}$. Finally, property 1. also holds for $P_{(x_1,\ldots,x_q;y_1,\ldots,y_p)}$, provided one considers the conditional discrete theory (see Section 1.5.2).*

PROOF. Property 1 has been proved in Propositions 2.46 and 2.53, for the measure $P$ and for its conditional versions.

If $M$ is closed, property 2 is actually property 2 of Proposition 2.43. If $M$ has a boundary, consider a minimal closure $M_1$ of $M$, a path $c$ of $M$ and a sequence $(c_n)$ of paths on $M$ converging to $c$ with fixed endpoints. This convergence still holds if we regard our paths as paths on $M_1$, so that, for all $x,y$, $H_{c_n}$ converges in probability towards $H_c$ under $P^{M_1}_{(x_1,\ldots,x_1;y_1,\ldots,y_p)}$, hence under $P_{(x_1,\ldots,x_1;y_1,\ldots,y_p)}$. Now, since $G$ is bounded, the distance between two variables is bounded above, and we can integrate this convergence with respect to $x$ and $y$, and finally find that it holds for $P$ and all its conditional versions.

By Lemma 2.51, properties 1 and 2 determine the distribution of any finite family $(H_{c_1},\ldots,H_{c_n})$ under $P$ (or one of its conditional versions, see property 6). Since the $\sigma$-field is the cylinder $\sigma$-field, this actually determines the full measure.

If $M$ is closed, property 3 is property 5 of Proposition 2.43. If $M$ has a boundary, then, by construction, property 3 is true on $M$ under $P_{(x_1,\ldots,x_1;y_1,\ldots,y_p)}$ because it is true on a minimal closure. Here also, we can integrate this almost sure equality with respect to $x$ and $y$ and show that Property 3. holds under $P$ and all its conditional versions.

Let $\psi$ be an area-preserving diffeomorphism. Because of the property 1 and of Proposition 1.27, the equality in distribution $(H_{c_1},\ldots,H_{c_n}) \stackrel{(d)}{=} (H_{\psi(c_1)},\ldots,H_{\psi(c_n)})$ holds for any finite family $(c_1,\ldots,c_n)$ which can be realized as a family of paths in a graph. By Lemma 2.51 again and property 2, this equality actually holds for any finite family and property 4 is proved.

Property 5 has been proved in Propositions 2.52 and 2.57. □

As explained in the introduction, the general purpose of this work, and more generally of stochastic quantization of Yang-Mills theory, is to construct a probability measure on the space of connections modulo gauge transformations on a principal bundle over a compact surface. The space of test functions or observables we are interested in is the space of gauge-invariant functions of the holonomy of a connection along a finite family of loops.

This is why we have sought to construct a random holonomy rather than a random connection, and indeed that is what Theorem 2.58 provides us with, under the form of the distribution of a $G$-valued stochastic process indexed by the set of piecewise embedded paths on $M$. Yet we have noticed earlier that a deterministic connection *modulo gauge transformations* does not induce a well-defined parallel transport along each path of $PM$, at least not as an element of $G$ (see Equation (1.2)). Along an open path for example, it determines *nothing* intrinsically, and along a loop, it determines a conjugacy class of $G$.

This gives geometrical evidence that there is, in a sense, too much information in the probability space $(\mathcal{F}(PM,G),\mathcal{C},P)$ given by Theorem 2.58, and that understanding this point will involve gauge transformations.

Before investigating the role of gauge transformations, let us discuss the existence of a multiplicative version for the random holonomy process.

**2.10.2. Multiplicative random holonomy.** So far, pathwise regularity properties for the random holonomy process have stayed out of reach, because, among other things, of the unusual size of the index set, which is larger than what traditionnal tools can handle. In this section, we are going to prove that there exists a *multiplicative* version of the Yang-Mills measure. This means that it is possible to realize the measure on the space of multiplicative maps from $PM$ to $G$. This improvement will for example allow us to understand much better the role of gauge transformations.

Set

$$\mathcal{M}(PM,G) = \{f \in \mathcal{F}(PM,G) : \forall\, c,c_1,c_2 \in PM, c \simeq c_1 c_2 \Rightarrow f(c) = f(c_2)f(c_1)\}.$$

As a subset of $\mathcal{F}(PM, G)$ endowed with the product topology, $\mathcal{M}(PM, G)$ is a compact Hausdorff topological space. Its Borel $\sigma$-field coincides with its cylinder $\sigma$-field that we denote by $\mathcal{C}$. In order to put a probability measure on $(\mathcal{M}(PM, G), \mathcal{C})$, we shall regard this measurable space as the projective limit of a family of measurable spaces.

Let $\Lambda$ denote the set of finite subsets of $PM$. When ordered by inclusion, $\Lambda$ becomes a directed set (see Definition 2.1). In what follows, we use the results of Theorem 2.58 without explicit reference.

Let $I$ be an element of $\Lambda$. The distribution of $(H_c)_{c \in I}$, that is, the finite-dimensional marginal of the Yang-Mills measure corresponding to $I$, determines a Borel probability measure $P^I$ on $G^I$. In order to incorporate mutliplicativity at this level already, let us define

$$\Omega_I = \{g = (g_c)_{c \in I} \in G^I : \forall\, c, c_1, c_2 \in I, c \simeq c_1 c_2 \Rightarrow g_c = g_{c_2} g_{c_1}\}.$$

Observe that $\Omega_I$ is a compact space whose Borel $\sigma$-field coincides with the trace of $\mathcal{B}(G^I)$. The multiplicativity of the random holonomy tells us exactly that $\Omega_I$ contains the support of $P^I$.

The same argument works for any conditional version of the Yang-Mills measure: with the notation of Theorem 2.58 and for all $x \in G^q$, $y \in G^r$, $I \in \Lambda$, the distribution of $(H_c)_{c \in I}$ under $P_{x;y}$ determines a probability measure $P^I_{x;y}$ on $\Omega_I$.

Given $I$ and $J$ in $\Lambda$ such that $I \subset J$, the restriction mapping $f_{IJ} : G^J \longrightarrow G^I$ sends continuously $\Omega_J$ into $\Omega_I$ and one has $(f_{IJ})_* P^J_{x;y} = P^I_{x;y}$. Thus, for all $x, y$, the family of probability spaces $(\Omega_I, \mathcal{B}(\Omega_I), P^I_{x;y})$ and the mappings $f_{IJ}$ form a projective system in the sense of Definition 2.1. Now set

$$\Omega_{\mathcal{M}} = \varprojlim \Omega_I, \ I \in \Lambda.$$

Recall that for each $I \in \Lambda$, there is a natural mapping $f_I : \Omega_{\mathcal{M}} \longrightarrow \Omega_I$, and that $\Omega_{\mathcal{M}}$ is endowed with the $\sigma$-field $\mathcal{B}_{\mathcal{M}}$ generated by these mappings. Theorem 2.2 applies in this situation and gives us the following result.

PROPOSITION 2.59. *For all $x, y$, there exists on $(\Omega_{\mathcal{M}}, \mathcal{B}_{\mathcal{M}})$ a probability measure $P^{\mathcal{M}}_{x;y}$ such that for all $I \in \Lambda$, $(f_I)_* P^{\mathcal{M}}_{x;y} = P^I_{x;y}$. Moreover, $\mathcal{B}_{\mathcal{M}}$ is the Borel $\sigma$-field of the compact Hausdorff space $\Omega_{\mathcal{M}}$.*

It remains to identify the measurable space $(\Omega_{\mathcal{M}}, \mathcal{B}_{\mathcal{M}})$. We do this by identifying the topological space $\Omega_{\mathcal{M}}$.

PROPOSITION 2.60. *The mapping*

$$\begin{array}{rcl} \Omega_{\mathcal{M}} & \longrightarrow & \mathcal{M}(PM, G) \\ \omega = (\omega_I)_{I \in \Lambda} & \longmapsto & (c \mapsto \omega_{\{c\}}) \end{array}$$

*is a homeomorphism.*

PROOF. Let us denote by $F$ this mapping. It is continuous, because for each path $c$, $H_c \circ F = f_{\{c\}}$ is continuous on $\Omega_{\mathcal{M}}$, where $H_c$ denotes the coordinate mapping on $\mathcal{M}(PM, G)$. The inverse mapping is given by $F^{-1}(c \mapsto g_c) = ((g_c)_{c \in I})_{I \in \Lambda}$. Since both space are compact Hausdorff, $F$ is a homeomorphism. □

REMARK 2.61. Observe that Theorem 2.2 also asserts that the mappings $f_I : \Omega_{\mathcal{M}} \longrightarrow \Omega_I$ are onto for all $I$. In light of the identification just stated, this means that any $G$-valued multiplicative map defined on a finite subset of $PM$ (or actually any subset, as an easy extra argument shows) can be extended to a multiplicative map on the whole $PM$.

In the following theorem, we use the notation of Theorem 2.58.

THEOREM 2.62. *For all disintegration family, all $x \in G^q$, $y \in G^r$, there exists on $(\mathcal{M}(PM, G), \mathcal{C})$ a probability measure $P_{x;y}$ such that the distribution of the canonical process under $P_{x;y}$ is the corresponding conditional Yang-Mills measure. All the properties stated in Theorem 2.58 remain true.*

PROOF. There is almost nothing to prove. Observe only that $\mathcal{C}$ is the Borel $\sigma$-field of $\mathcal{M}(PG, M)$ so that we can transport directly the measures from $(\Omega_M, \mathcal{B}_M)$ to $(\mathcal{M}(PM, G), \mathcal{C})$ by using Proposition 2.60. □

It may be surprising to see that we have got a multiplicative version of the random holonomy so easily. What makes this construction work is the compacity of $G$ which has been used crucially in Theorem 2.2.

**2.10.3. Gauge transformations.** In this section, we want to investigate the action of gauge transformations on the probability space $(\mathcal{M}(PM, G), \mathcal{C}, P)$.

Given a graph on $M$, a discrete gauge transformation has been defined as an assignation of an element of $G$ to each vertex of the graph (see Definition 1.12). We extend this definition in the following way.

DEFINITION 2.63. By a *symbolic gauge transformation* we mean an element of the group $\mathcal{F}(M, G)$, that is, a mapping from $M$ to $G$. A symbolic gauge transformation $\phi$ acts on the set $\mathcal{M}(PM, G)$ according to the following rule:

$$(\phi \cdot \omega)(c) = \phi_{c(1)}^{-1} \omega(c) \phi_{c(0)} \, , \quad \omega \in \mathcal{M}(PM, G) \, , \, c \in PM. \tag{2.6}$$

It is readily checked that the multiplicativity is indeed preserved by this action. In the next proposition, we are using some notation about conjugacy classes of $G$ defined at the end of Section 1.5.3.

PROPOSITION 2.64. *Let $\phi$ be a symbolic gauge transformation. Let $L_1, \ldots, L_q$ be a disintegration family and $N_1, \ldots, N_r$ be boundary components of $M$. Let $m_i$ denote the basepoint of $L_i$ for each $i = 1, \ldots, q$ and $n_j$ that of $N_j$ for $j = 1, \ldots, r$. Then the action of $\phi$ on $(\mathcal{M}(PM, G), \mathcal{C})$ is measurable and, for any $x_1, \ldots, x_q$ and $y_1, \ldots, y_r$, we have*

$$\phi_* P_{(x_1, \ldots, x_q; y_1, \ldots, y_r)} = P_{(\mathrm{Ad}(\phi_{m_1})x_1, \ldots, \mathrm{Ad}(\phi_{m_q})x_q; \mathrm{Ad}(\phi_{n_1})y_1, \ldots, \mathrm{Ad}(\phi_{n_r})y_r)}. \tag{2.7}$$

*In particular, if $q = r = 0$, we have $\phi_* P = P$.*

*Moreover, given two collections $\mathfrak{x} = (\mathfrak{x}_1, \ldots, \mathfrak{x}_q)$ and $\mathfrak{y} = (\mathfrak{y}_1, \ldots, \mathfrak{y}_r)$ of conjugacy classes of $G$, the measure $P_{\mathfrak{x};\mathfrak{y}}$ defined by*

$$\begin{aligned} P_{\mathfrak{x};\mathfrak{y}} &= P_{(\mathfrak{x}_1, \ldots, \mathfrak{x}_q; \mathfrak{y}_1, \ldots, \mathfrak{y}_r)} \\ &= \int_{G^{q+r}} P_{(x_1, \ldots, x_q; y_1, \ldots, y_r)} \, d\delta_{(\mathfrak{x}_1, \ldots, \mathfrak{x}_q)}(x_1, \ldots, x_q) d\delta_{(\mathfrak{y}_1, \ldots, \mathfrak{y}_r)}(y_1, \ldots, y_q) \end{aligned}$$

*is invariant under the action of $\phi$. In other words, $\phi_* P_{\mathfrak{x};\mathfrak{y}} = P_{\mathfrak{x};\mathfrak{y}}$.*

PROOF. The second half of the statement follows easily from the first one, along the lines of Section 1.5.3.

In order to prove that (2.7) holds, we must prove that, for any finite family $(c_1, \ldots, c_n)$ of paths and any continuous function $f$ on $G^n$, the integrals of the function $f(H_{c_1}, \ldots, H_{c_n})$ on $\mathcal{M}(PM, G)$ against both measures are equal. This is equivalent to proving that

$$\int_{\mathcal{M}(PM,G)} f(\mathrm{Ad}_{\phi_{c_1(0)}^{-1}} H_{c_1}, \ldots, \mathrm{Ad}_{\phi_{c_n(0)}^{-1}} H_{c_n}) \, dP_{(x_1,\ldots,x_q;y_1,\ldots,y_r)} =$$
$$= \int_{\mathcal{M}(PM,G)} f(H_{c_1}, \ldots, H_{c_n}) \, dP_{(\mathrm{Ad}(\phi_{m_1})x_1,\ldots,\mathrm{Ad}(\phi_{m_q})x_q;\mathrm{Ad}(\phi_{n_1})y_1,\ldots,\mathrm{Ad}(\phi_{n_r})y_r)}.$$

By consistency with the discrete theory and Proposition 1.18, this holds whenever there exists a graph $\Gamma$ such that $L_1, \ldots, L_q$ and $c_1, \ldots, c_n$ belong to $\Gamma^*$.

Now, by the continuity property of the random holonomy (Theorem 2.58, Property 2.) and Lemma 2.51, we can extend this result to an arbitrary class of paths and the result is proved. $\square$

We use the adjective *symbolic* to qualify the elements of $\mathcal{F}(M, G)$ because they are, from the geometric point of view, the roughest possible approximation of what a smooth gauge transformation is. That Proposition 2.64 holds with such a rough gauge group indicates that we have lost a lot of topological information during the construction.

Notice that, just as in Section 1.8.1, the invariance under symbolic gauge transformations implies that the distribution of the random holonomy along any open path under $P_{\mathfrak{x}}$ is uniform. This is again a probabilistic indication that there is too much information in our probability space. It is time now to give a precise content to this affirmation.

**2.10.4. Invariant events.** Let us define the gauge-invariant $\sigma$-field $\mathcal{I}$ on $\mathcal{M}(PM, G)$ as
$$\mathcal{I} = \{A \in \mathcal{C} : \phi \cdot A = A \; \forall \phi \in \mathcal{F}(M, G)\}.$$
Our aim in this section is to describe $\mathcal{I}$. Since $\mathcal{C}$ is generated by cylinders, it is natural to begin our investigation by a description of the invariant cylinders. Actually, let us start by providing some examples of invariant cylinders.

Recall that a symbolic gauge transformation acts on an element of $\mathcal{M}(PM, G)$ in particular by conjugating its value on loops, more precisely by conjugating by the same element of $G$ its value on all loops based at the same point. Let us define the *joint conjugacy classes* in $G^n$ as the orbits of the diagonal action of $G$ by adjunction on $G^n$. In other words, $(x_1, \ldots, x_n)$ and $(y_1, \ldots, y_n)$ are in the same joint conjugacy class if and only if there exists $g$ in $G$ such that, for all $i = 1, \ldots, n$, $y_i = \mathrm{Ad}(g)x_i$. We denote by $G^n/\mathrm{Ad}$ the space of joint conjugacy classes in $G^n$ and denote the class of $(x_1, \ldots, x_n)$ by $[x_1, \ldots, x_n]$. Observe that $[x_1, \ldots, x_n]$ determines the usual conjugacy classes $\mathfrak{x}_1, \ldots, \mathfrak{x}_n$ but that the converse is false in general. Each element of $G^n/\mathrm{Ad}$ is a compact subset of $G^n$, so that $G^n/\mathrm{Ad}$ can be endowed with the Hausdorff distance between compact subsets defined by

$$d(K_1, K_2) = \sup(\sup_{k_1 \in K_1} \inf_{k_2 \in K_2} d(k_1, k_2), \sup_{k_2 \in K_2} \inf_{k_1 \in K_1} d(k_1, k_2)).$$

Using the invariance by adjunction of the distance in $G$, one easily checks that the canonical projection $G^n \longrightarrow G^n/\mathrm{Ad}$ is 1-Lipschitz continuous. In particular, $G^n/\mathrm{Ad}$ endowed with the distance $d$ is a compact space.

Let $(H_c)_{c\in PM}$ denote the canonical process on $\mathcal{M}(PM,G)$ and let us define, for any point $m$ on $M$ and any finite family $l_1,\ldots,l_n$ of loops based at $m$ the $G^n/\mathrm{Ad}$-valued random variable
$$\mathfrak{H}_{l_1,\ldots,l_n} = [H_{l_1},\ldots,H_{l_n}].$$
It is an example of a gauge-invariant cylindrical random variable and the next result says that there are essentially no others. We denote by $L_m M$ the set of loops based at $m$.

PROPOSITION 2.65. *The $\sigma$-field of gauge-invariant events is generated by the joint conjugacy classes of random holonomies along finite families of loops based at the same point. For short,*
$$\mathcal{I} = \sigma(\mathfrak{H}_{l_1,\ldots,l_n} : l_1,\ldots,l_n \in L_m M, m \in M, n \geq 1).$$

Let $\mathcal{L}$ denote the $\sigma$-field on the right-hand side of the equality. It is clear that $\mathcal{I} \supset \mathcal{L}$. We are going to prove the other inclusion by using Blackwell's theorem (see [18]) but this cannot be done directly, for Blackwell's theorem applies only to separable $\sigma$-fields.

Let $S$ be a subset of $PM$. We say that an event $A \in \mathcal{C}$ is an $S$-cylinder if there exists a Borel subset $A_S$ of $G^S$ such that
$$A = \{\omega \in \mathcal{M}(PG,M) : (\omega(c))_{c\in S} \in A_S\}.$$
Recall that any event of $\mathcal{C}$ is an $S$-cylinder for some *countable* set $S$.

Using a notation similar to that used for graphs in Chapter 1, we denote by $S^*$ the set of paths which can be obtained by concatenation of a finite number of paths of $S$. If $S$ is countable, then so is $S^*$. Any $S$-cylinder is clearly an $S^*$-cylinder, just because $S \subset S^*$. Moreover, the restriction of a multiplicative function to $S^*$ is completely determined by its restriction to $S$, so that an $S^*$-cylinder is also an $S$-cylinder. Blackwell's theorem will allow us to prove the following result which implies immediately Proposition 2.65.

PROPOSITION 2.66. *Let $S$ be a countable subset of $PM$. Let $\mathcal{C}_S = \sigma(H_c, c \in S)$ denote the $\sigma$-field of $S$-cylinders, $\mathcal{I}_S = \mathcal{I} \cap \mathcal{C}_S$ that of invariant $S$-cylinders. Let $\mathcal{L}_S = \sigma(\mathfrak{H}_{l_1,\ldots,l_n} : l_i \in S^* \cap LM)$. Then*
$$\mathcal{I}_S = \mathcal{L}_S.$$

LEMMA 2.67. *Let $S$ be any subset of $PM$. Let $\omega_1$ and $\omega_2$ be two elements of $\mathcal{M}(PM,G)$ and assume that for all $m \in M$, $n \geq 1$ and all $l_1,\ldots,l_n \in S^* \cap L_m M$,*
$$[\omega_1(l_1),\ldots,\omega_1(l_n)] = [\omega_2(l_1),\ldots,\omega_2(l_n)]. \tag{2.8}$$
*Then there exists a symbolic gauge transformation $\phi$ such that $\omega_1$ and $\phi \cdot \omega_2$ agree on $S^*$.*

The proof is inspired by that of Proposition 2.1.2 in [36].

PROOF. Let $S$, $\omega_1$ and $\omega_2$ be as in the statement. Let $V_S$ denote the set of all endpoints of the paths of $S$. Let us assume that the graph $(V_S, S)$ is connected in the usual graph-theoritic sense, that is, that any two points of $V_S$ can be joined by a path of $S^*$.

Let $v$ be a fixed point of $V_S$. For each loop $l$ based at $v$ and belonging to $S^*$, let $K_l$ denote the subset of those $g \in G$ for which $\omega_1(l) = \mathrm{Ad}(g)\omega_2(l)$. It is a compact subset of $G$, and, by assumption (2.8), $K_{l_1} \cap \ldots \cap K_{l_n}$ is non-empty for every finite family $l_1, \ldots, l_n$. Hence, there exists $g_v \in \bigcap K_l$ and it satisfies $\omega_1(l) = \mathrm{Ad}(g_v)\omega_2(l)$ for all $l \in S^* \cap LM$.

By replacing $\omega_2$ by $\phi \cdot \omega_2$, where $\phi_m = g_v$ if $m = v$ and 1 otherwise, we can assume that $\omega_1(l) = \omega_2(l)$ for all loop $l \in S^* \cap LM$. Now let $w$ be a point of $V_S$. If $c \in S^*$ is a path from $v$ to $w$, then the quantity $\omega_2(c)\omega_1(c)^{-1}$ depends only on $w$, because if $c'$ is another path from $v$ to $w$, then $c'c^{-1}$ is a loop based at $v$ and $\omega_1(c)^{-1}\omega_1(c') = \omega_2(c)^{-1}\omega_2(c')$. Hence we can put $\phi_w = \omega_2(c)\omega_1(c)^{-1}$ and this defines a $G$-valued function on $V_S$. Extending this function arbitrarily outside $V_S$, we get a symbolic gauge transformation which is readily checked to be such that $\omega_1$ and $\phi \cdot \omega_2$ agree on $S^*$.

If the graph $(V_S, S)$ is not connected, one can do this construction on each connected component, thus defining $\phi$ separately on each component of $V_S$. □

PROOF OF PROPOSITION 2.66. The separated measurable space associated to $(\mathcal{M}(PM, G), \mathcal{C}_S)$ is isomorphic to $(\mathcal{M}(S, G), \sigma(H_c, c \in S))$, which is a metrizable compact Borel space, so that $(\mathcal{M}(PM, G), \mathcal{C}_S)$ is a Blackwell space. The $\sigma$-field $\mathcal{L}_S$ is separable because it is generated by a countable set of random variables. Hence, by Blackwell's theorem, $\mathcal{I}_S \subset \mathcal{L}_S$ holds if and only if every atom of $\mathcal{I}_S$ is a union of atoms of $\mathcal{L}_S$.

Let $\omega_1$ and $\omega_2$ be two functions of $\mathcal{M}(PM, G)$ that belong to the same atom of $\mathcal{L}_S$. This means exactly that they satisfy assumption (2.8). Hence there exists a symbolic gauge transformation $\phi$ such that $\omega_1$ and $\phi \cdot \omega_2$ agree on $S^*$. Now any invariant $S$-cylinder containing $\omega_1$ contains also $\phi \cdot \omega_2$, hence $\omega_2$. In other words, $\omega_1$ and $\omega_2$ belong to the same atom of $\mathcal{I}_S$. Blackwell's theorem implies that $\mathcal{I}_S \subset \mathcal{L}_S$ and the other inclusion is obvious. □

REMARK 2.68. If $S$ is a countable subset of $L_m M$ for some $m \in M$, then the random variable $\mathfrak{H}_S = [(H_l)_{l \in S}]$ is not only a $G^S/\mathrm{Ad}$-valued random variable, but in fact a $\mathcal{M}(S, G)/\mathrm{Ad}$-valued random variable. By $\mathcal{M}(S, G)$ we mean of course the set of multiplicative $G$-valued maps defined on $S$, which is a stable subset of $G^S$ for the diagonal adjoint action.

It is actually possible to improve Proposition 2.65 in many cases, depending on the nature of $G$. In [37], A. Sengupta has established the group-theoretic result which underpins the following statement.

PROPOSITION 2.69. *Let $S$ be a subset of $LM$. Let $\omega_1$ and $\omega_2$ be two multiplicative functions such that, for all $l \in S^*$,*

$$[\omega_1(l)] = [\omega_2(l)].$$

*If $G$ is a product of groups of this list : Abelian, $U(n)$, $SU(n)$, $SO(2n+1)$, or one of the following groups : $Spin(2n+1)$, $Pin(n)$, $SO(4)$, $Spin(4)$, then*

$$[\omega_1(l_1), \ldots, \omega_1(l_n)] = [\omega_2(l_1), \ldots, \omega_2(l_n)]$$

*for all $l_1, \ldots, l_n \in S^*$.*

Then a slight modification in the proof of Proposition 2.66 shows that the following result holds.

PROPOSITION 2.70. *Under the assumption on $G$ made in Proposition 2.69, one has the equality*
$$\mathcal{I} = \sigma(\mathfrak{H}_l : l \in LM).$$

**2.10.5. The Yang-Mills process.** In the previous section, we have identified the sub-$\sigma$-field of $\mathcal{C}$ which contains the correct amount of information as far as gauge-invariant random variables are concerned. We expect a result very close to Theorem 2.62 to be true, when $\mathcal{C}$ replaced by $\mathcal{I}$. This actually requires some adjustments.

The measure $P$ and its conditional versions $P_{x;y}$ are of course well-defined by restriction on $(\mathcal{M}(PM, G), \mathcal{I})$, but one should observe that the random variable $(H_{L_1}, \ldots, H_{L_q}, H_{N_1}, \ldots, H_{N_r})$, with respect to which the measures $P_{x;y}$ are supposed to disintegrate $P$, is not $\mathcal{I}$-measurable. Since we are now working with an invariant $\sigma$-field, Proposition 2.64 implies that $P_{x;y}$ depends on $x$ and $y$ only through the conjugacy classes $\mathfrak{x} = (\mathfrak{x}_1, \ldots, \mathfrak{x}_q) \in (G/\mathrm{Ad})^q$ and $\mathfrak{y} = (\mathfrak{y}_1, \ldots, \mathfrak{y}_r) \in (G/\mathrm{Ad})^r$, and that $P_{x;y} = P_{\mathfrak{x};\mathfrak{y}}$.

On the other hand, if $\eta(x, y)$ denotes the distribution of the random variable $(H_{L_1}, \ldots, H_{L_q}, H_{N_1}, \ldots, H_{N_r})$ under $P$ seen as a measure on $\mathcal{C}$, then we know by property 5 of Theorem 2.58 that the equality of measures $P = \int_{G^{q+r}} P_{x;y} \, d\eta(x,y)$ holds. If we restrict this equality to $\mathcal{I}$, we find $P = \int_{(G/\mathrm{Ad})^{q+r}} P_{\mathfrak{x};\mathfrak{y}} \, d\bar{\eta}(\mathfrak{x}, \mathfrak{y})$, where $\bar{\eta}$ is the projection on $(G/\mathrm{Ad})^{q+r}$ of $\eta$, that is, the distribution of the $\mathcal{I}$-measurable random variable $(\mathfrak{H}_{L_1}, \ldots, \mathfrak{H}_{L_q}, \mathfrak{H}_{N_1}, \ldots, \mathfrak{H}_{N_r})$. This proves that a disintegration property still holds.

The consistency with the discrete theory is clearly preserved by reducing the $\sigma$-field. Continuity properties are also preserved, since we have endowed $G^n/\mathrm{Ad}$ with such a distance that the canonical projection $G^n \longrightarrow G^n/\mathrm{Ad}$ is 1-Lipschitz continuous (see Section 2.10.4). The distance that we use between $G^n/\mathrm{Ad}$-valued random variable is of course again the distance in probability
$$d_P(X,Y) = \mathbb{E}\, d(X,Y).$$

What may look more worrying is that the multiplicativity hardly makes sense with conjugacy classes. Let us state a simple lemma which explains how one can multiply *joint* conjugacy classes. If $S$ is a subset of $LM$, recall that the restriction to $S$ of a multiplicative $G$-valued function determines its restriction to $S^*$.

LEMMA 2.71. *Let $S$ be a subset of $L_m M$ for some $m$ in $M$. The mapping $\mathcal{M}(S, G) \longrightarrow \mathcal{M}(S^*, G)$ is compatible with the diagonal adjoint action of $G$ and it determines a quotient mapping*
$$m_S : \mathcal{M}(S, G)/\mathrm{Ad} \longrightarrow \mathcal{M}(S^*, G)/\mathrm{Ad}.$$

The proof is left to the reader. We are finally able to state the following theorem, which is written as to be as self-contained as possible.

THEOREM 2.72. *Let $(\mathcal{M}(PM, G), \mathcal{I})$ be the space of multiplicative maps from $PM$ to $G$ endowed with the gauge-invariant cylindrical $\sigma$-field. For all finite family $l_1, \ldots, l_n$ of loops based at the same point in $M$, the random variable $\mathfrak{H}_{l_1,\ldots,l_n} = [H_{l_1}, \ldots, H_{l_n}]$ is $\mathcal{I}$-measurable and depends on $l_1, \ldots, l_n$ only through their equivalence classes (see Definition 2.12). These variables generate $\mathcal{I}$ (see Section 2.10.4).*

*They also satisfy the following multiplicativity property. If $S$ is a finite or countable subset of $LM$, then $\mathfrak{H}_{S^*} = m_S \circ \mathfrak{H}_S$ (see Lemma 2.71).*

*There exists on* $(\mathcal{M}(PM,G),\mathcal{I})$ *a probability measure* $P$ *called the Yang-Mills measure which has the following properties.*

1. *Let* $l_1,\ldots,l_n$ *be a finite family of loops based at the same point. If there exists a graph, say* $\Gamma$, *such that* $l_1,\ldots,l_n$ *can be written as concatenations of edges of* $\Gamma$, *then the distribution of* $\mathfrak{H}_{l_1,\ldots,l_n}$ *under* $P$ *is equal to that of* $[h_{l_1},\ldots,h_{l_n}]$ *under* $P^\Gamma$ *(see Section 1.4).*

2. *Let* $((l_{1,k},\ldots,l_{n,k}))_{k\geq 0}$ *be a sequence of* $n$*-tuples of loops such that, for each* $k \geq 0$, *the loops* $l_{1,k},\ldots,l_{n,k}$ *are based at the same point and, for each* $i = 1,\ldots,n$, *there exists a loop* $l_i$ *such that* $l_{i,k}$ *converges in length to* $l_i$ *as* $k$ *tends to infinity. Then one has the convergence in probability*

$$\mathfrak{H}_{l_{1,k},\ldots,l_{n,k}} \xrightarrow[k\to\infty]{P} \mathfrak{H}_{l_1,\ldots,l_n}.$$

*The measure* $P$ *is characterized by these two properties. It also satisfies the following other properties.*

3. *It is invariant under area-preserving diffeomorphism, in the same sense as stated in Property 4 of Theorem 2.58.*

*Let* $L_1,\ldots,L_q$ *be a disintegration family and* $N_1,\ldots,N_r$ *be boundary components of* $M$. *For all* $\mathfrak{x} \in (G\mathrm{Ad})^q$, $\mathfrak{y} \in (G/\mathrm{Ad})^r$, *there is a probability measure* $P_{\mathfrak{x};\mathfrak{y}}$ *on* $(\mathcal{M}(PM,G),\mathcal{I})$ *such that the following properties hold.*

4. *The mapping* $(\mathfrak{x},\mathfrak{y}) \mapsto P_{\mathfrak{x};\mathfrak{y}}$ *disintegrates the Yang-Mills measure* $P$ *with respect to the random variable* $(\mathfrak{H}_{L_1},\ldots,\mathfrak{H}_{L_q},\mathfrak{H}_{N_1},\ldots,\mathfrak{H}_{N_r})$.

5. *Properties 1 and 2 are true when* $P$ *is replaced with* $P_{\mathfrak{x};\mathfrak{y}}$, *provided one considers conditional discrete measures in Property 2.*

The space of based loops will play an important role in the next chapters. The following simple observation will prove very useful.

PROPOSITION 2.73. *Let* $m$ *be a fixed point of* $M$. *The family* $(\mathfrak{H}_{l_1,\ldots,l_n})_{l_i \in L_m M}$ *generates the* $\sigma$*-algebra* $\mathcal{I}$.

PROOF. Let $l_1,\ldots,l_n$ be $n$ loops on $M$ based at a point $m_1$. Let $c$ be an arbitrary path joining $m$ to $m_1$. Then the equality

$$\mathfrak{H}_{cl_1c^{-1},\ldots,cl_nc^{-1}} = \mathfrak{H}_{l_1,\ldots,l_n}$$

proves the result. $\square$

CHAPTER 3

# Abelian gauge theory

In this chapter, we continue the investigation of the Abelian case $G = U(1)$ started in Section 1.9. Recall that we have chosen to work with deterministic boundary conditions when $M$ has a boundary. The key results obtained in the discrete setting are Proposition 1.36, which says in particular that the random holonomies along a system of generators of $H_1(M)$ are independent, uniformly distributed, and independent of the holonomies along homologically trivial cycles, and Proposition 1.43 which describes how one can realize the distribution of the random holonomy along a family of homologically trivial loops by using their double layer potential and a white noise of intensity $\sigma$ on $M$.

The first part of this chapter will be devoted to the extension of these results to the continuous setting. This will be quite easy with the results of Chapter 2 in hand, in particular Corollary 2.45 which says that the double layer potential of a loop depends continuously on this loop, in the appropriate topologies. This extension will allow us to give an alternative description of the distribution of the Yang-Mills process based on a white noise on $M$.

Then we will show that it is possible to construct a white noise on $M$ on the same probability space where the Yang-Mills process lives, and that this white noise plays the role of the curvature of the random connection which is supposed to underpin the random holonomy process.

## 3.1. The random holonomy as a white noise functional

Let us begin by some remarks about the continuous Yang-Mills measure in the case $G = U(1)$. The first point is that the symbolic gauge transformations act trivially on $\mathcal{M}(PM, G)$: the random variables $H_l$ are now equal to their conjugacy class $\mathfrak{H}_l$ and the $\sigma$-field $\mathcal{I}$ is equal to $\sigma(H_l : l \in LM)$. Then, just as at the beginning of Section 1.9.1, we can extend the set of variables $(H_l)_{l \in LM}$ by multiplicativity to a set $(H_c)_{c \in CM}$ indexed by *cycles*, i.e. linear combinations of loops with integral coefficients. This family is more natural in this context because it allows us to use freely the homology of the surface $M$.

Recall that $M$ is endowed with a Riemannian metric, which allows us to define double layer potentials, and that we have chosen $2g$ piecewise geodesic loops $\ell_1, \ldots, \ell_{2g}$ such that $\mathcal{B} = ([\ell_1], \ldots, [\ell_{2g}], [N_1], \ldots, [N_{p-1}])$ is a basis of $H_1(M; \mathbb{Z})$. We have fixed $x_1, \ldots, x_p$ $p$ elements of $U(1)$ corresponding to the boundary conditions. Recall also that $W$ is a white noise on $M$ with intensity $\sigma$, that is, an isometry from $L^2(M, \sigma)$ into a subspace of some $L^2(\Omega, \mathcal{A}, \mathbb{P})$ consisting of centered Gaussian random variables. We denote by $W(u)$ the Gaussian variable associated to the square-integrable function $u$. Let us assume that the probability space $(\Omega, \mathcal{A}, \mathbb{P})$ also supports $2g$ independent variables $U_1, \ldots, U_{2g}$ uniformly distributed on $U(1)$, independent of $W$, and another variable $T$ independent of all these, with the distribution described in Proposition 1.38.

We want to construct on $(\Omega, \mathcal{A}, \mathbb{P})$ a family of random variables $(H_c^W)_{c \in CM}$ with the same distribution as $(H_c)_{c \in CM}$ under $P_{(x_1,\ldots,x_p)}$.

Let $c$ be a cycle in $CM$ and consider its decomposition
$$c = \lambda_1 \ell_1 + \ldots + \lambda_{2g} \ell_{2g} + \nu_1 N_1 + \ldots + \nu_{p-1} N_{p-1} + c^\perp,$$
where $\lambda_1, \ldots, \lambda_{2g}, \nu_1, \ldots, \nu_{p-1}$ are integers and $c^\perp$ belongs to the set $C_0\Gamma$ of cycles homologous to zero.

By multiplicativity, we need only to define $H_{\ell_i}^W$ for $i = 1 \ldots 2g$, $H_{N_j}^W$ for $j = 1 \ldots p-1$ and $H_{c^\perp}^W$. Clearly, we want $H_{N_j}^W$ to be equal to $x_j$ for all $j = 1 \ldots p-1$. Proposition 1.36 indicates that the random variables $H_{\ell_1}^W, \ldots, H_{\ell_{2g}}^W$ must be chosen uniform on $U(1)$ and independent of the random holonomy along any homologically trivial cycle. Let us define $H_{\ell_i}^W = U_i$ for each $i$ and there remains only to define $H_{c^\perp}^W$.

Let us consider the double layer potential $u_{c^\perp}$ of $c^\perp$ (see Definition 1.39) and the projection $W_0$ of $W$ on the hyperplane of functions with mean equal to zero: $W_0(u) = W(u - \frac{1}{\sigma(M)} \int_M u \, d\sigma)$. Recall that we are able to define the relative area $\sigma(c^\perp)$ enclosed by the homologically trivial cycle $c^\perp$ (see before Proposition 1.43), which belongs to $\mathbb{R}$ when $M$ has a boundary and $\mathbb{R}/\mathbb{Z}$ when $M$ is closed.

DEFINITION 3.1. For each cycle $c \in CM$ with the decomposition
$$c = \lambda_1 \ell_1 + \ldots + \lambda_{2g} \ell_{2g} + \nu_1 N_1 + \ldots + \nu_{p-1} N_{p-1} + c^\perp,$$
we define the following random variable in $L^\infty(\Omega, \mathcal{A}, \mathbb{P})$:
$$H_c^W = \exp 2i\pi (W_0(u_{c^\perp}) + \sigma(c^\perp)T) \, U_1^{\lambda_1} \ldots U_{2g}^{\lambda_{2g}} x_1^{\nu_1} \ldots x_{p-1}^{\nu_{p-1}}.$$

THEOREM 3.2. *The family of random variables $(H_c^W)_{c \in CM}$ has the same distribution as the family $(\mathfrak{H}_c)_{c \in CM}$ under $P_{(x_1,\ldots,x_p)}$.*

The proof of this theorem involves two steps : the identification of the distributions on the set of piecewise geodesic loops and then a continuity argument. The first part is easy, but the second requires the proof that the new family of variables satisfies some regularity property.

We begin by extending the distance $d_\ell$ to the space $CM$ of cycles.

DEFINITION 3.3. Let $c$ and $c'$ be two cycles on $M$. Set
$$d_\ell(c, c') = 1 \wedge \inf \sum_{i=1}^{k} n_i d_\ell(l_i, l_i'),$$
where the infimum is taken over the integers $k, n_1, \ldots, n_k$ and the loops $l_1, l_1', \ldots, l_k, l_k'$ such that $c = n_1 l_1 + \ldots + n_k l_k$ and $c' = n_1 l_1' + \ldots + n_k l_k'$.

The continuity property can be stated as follows.

PROPOSITION 3.4. *The mapping $c \mapsto H_c^W$ is continuous from $(CM, d_\ell)$ into $L^2(\Omega, \mathcal{A}, \mathbb{P})$.*

To begin with, observe that homology classes are both open and closed for $d_\ell$, so that $c \mapsto U_1^{\lambda_1} \ldots U_{2g}^{\lambda_{2g}} x_1^{\nu_1} \ldots x_{p-1}^{\nu_{p-1}}$ is locally constant on $(CM, d_\ell)$ and $c^\perp$ depends continuously on $c$.

Now the crucial point is the continuity of the mapping $c^\perp \mapsto u_{c^\perp}$. Proposition 2.45 applies to loops on a closed surface. The extension to cycles is trivial, but not that to surfaces with boundary. A topological description of the double layer potential of a cycle homologous to zero will be useful.

Let us fix $c \in C_0 M$, without making any assumption on the boundary of $M$. Let $x$ and $y$ be two points of $M$ outside the image of $c$. If necessary, we modify locally the $\ell_i$'s in a neighbourhood of $x$ and $y$ in order to make sure that $x$ nor $y$ meet any of them. Let $\ell_x$ be the boundary of a small disk $D_x$ around $x$, small enough not to meet $c$ and not to contain $y$. Define $\ell_y$ similarly. The module $H_1(M - \{x,y\})$ is generated by $[\ell_1], \ldots, [\ell_{2g}], [N_1], \ldots, [N_{p-1}], [\ell_x], [\ell_y]$. In $M - \{x, y\}$, we have the equality

$$[c] = \sum_{i=1}^{2g} \lambda_i [\ell_i] + \sum_{j=1}^{p-1} \nu_j [N_j] + p[\ell_x] + q[\ell_y]$$

for some $p, q \in \mathbb{Z}$. This equality also holds in $M$, where $[c] = [\ell_x] = [\ell_y] = 0$, and this proves that $\lambda_i = \nu_j = 0$. Thus, $[c] = p[\ell_x] + q[\ell_y]$ in $M - \{x, y\}$.

LEMMA 3.5. *With the same notation,* $u_c(x) - u_c(y) = p - q$.

PROOF. Recall that $G$ denotes the Green function on $M$. The 1-form $*dG_x - *dG_y$ is closed on $M - \{x, y\}$, so that

$$\begin{aligned}
u_c(x) - u_c(y) &= \int_c *dG_x - *dG_y \\
&= p \int_{\ell_x} *dG_x - *dG_y + q \int_{\ell_y} *dG_x - *dG_y \\
&= p(u_{\ell_x}(x) - u_{\ell_x}(y)) + q(u_{\ell_y}(x) - u_{\ell_y}(y)) \\
&= p \left( 1_{D_x}(x) - \frac{\sigma(D_x)}{\sigma(M)} - 1_{D_x}(y) + \frac{\sigma(D_x)}{\sigma(M)} \right) + q \left( 1_{D_y}(x) - 1_{D_y}(y) \right) \\
&= p - q.
\end{aligned}$$
□

REMARK 3.6. If $M$ is closed, then $[\ell_x] + [\ell_y] = 0$ in $H_1(M - \{x, y\})$. In this case, $p$ and $q$ are defined only up to addition of a constant. However, $p - q$ is well-defined.

COROLLARY 3.7. *Assume $M$ has a boundary. If $M_1$ is a minimal closure of $M$ and if we identify $M$ with a submanifold of $M_1$, then for all $c \in C_0 M$, the potentials $u_c^M$ and $u_c^{M_1}|_M$ computed respectively in $M$ and $M_1$ differ only by an additive constant.*

PROOF. Both functions $u_c^M$ and $u_c^{M_1}$ are locally constant on the complement of the range of $c$. Moreover, if the equality $[c] = p[\ell_x] + q[\ell_y]$ holds in $H_1(M - \{x, y\})$, then it also holds in $H_1(M_1 - \{x, y\})$, so that for all $x, y$ outside the range of $c$, $u_c^M(x) - u_c^M(y) = u_c^{M_1}(x) - u_c^{M_1}(y)$. □

COROLLARY 3.8. *The map $c \mapsto W_0(u_{c^\perp})$ is continuous.*

PROOF. If $M$ is closed, this is a consequence of Proposition 2.45 and the fact that $c^\perp$ depends continuously on $c$.

If $M$ has a boundary, let us fix a minimal closure $M_1$ of $M$. The preceding corollary tells us that

$$u_{c^\perp}^M - \frac{1}{\sigma(M)} \int_M u_{c^\perp}^M \, d\sigma = u_{c^\perp}^{M_1}{}_{|M} - \frac{1}{\sigma(M)} \int_M u_{c^\perp}^{M_1} \, d\sigma.$$

Since we use the centered white noise $W_0$, this terminates the proof. □

Finally, let us study the term $\sigma(c)$, when $c \in C_0 M$. We will show that it can be extracted from the double layer potential of $c$.

LEMMA 3.9. *Let $c$ be a cycle of $C_0 M$.*
*1. If $M$ has no boundary, $\sigma(c)$ is the element $t \in \mathbb{R}/\mathbb{Z}$ such that $u_c$ takes its values in $\mathbb{Z} - t$.*
*2. If $M$ has a boundary, consider $M_1$ a minimal closure of $M$. Then $\sigma(c)$ is equal to $-\frac{\sigma(M_1)}{\sigma(M)}$ times the value of $u_c^{M_1}$ at any point of $M_1 - M$.*

PROOF. 1. Let $x$ be a point of $M$ outside the image of $c$. Recall that $\sigma(c)$ is defined as the class modulo 1 of the integral $\frac{1}{\sigma(M)} \int_c \gamma$, where $\gamma \in \Omega^1(M - \{x\})$ satisfies $d\gamma = \sigma$. This class depends neither on $x$ nor on $\gamma$. For example, $\gamma = -\sigma(M) \cdot *dG_x$ is a possible choice. With this choice, it becomes clear that

$$u_c(x) = -\sigma(c) \pmod{1}.$$

2. Let $x$ be a point of $M_1 - M$. The form $*dG_x^{M_1}$ is smooth on $M$ and satisfies $d * dG_x^{M_1} = -\frac{1}{\sigma(M_1)} \sigma$. Hence,

$$u_c^{M_1}(x) = \int_c *dG_x^{M_1} = -\frac{\sigma(M)}{\sigma(M_1)} \sigma(c).$$

□

COROLLARY 3.10. *The map $c \mapsto \sigma(c)$ defined on $C_0 M$ is continuous.*

PROOF. Let $(c_n)_{n \geq 0}$ be a sequence of cycles of $C_0 M$ such that $c_n \xrightarrow{\ell} c$. Since $u_{c_n} \xrightarrow{L^2} u_c$ and since all these functions are locally constant, there is pointwise convergence outside the image of $c$. The result follows. □

This corollary finishes the proof of Proposition 3.4. We shall need a last property, the multiplicativity of the family $(H_c^W : c \in CM)$.

PROPOSITION 3.11. *For any cycles $c_1$ and $c_2$ in $CM$,*

$$H_{c_1+c_2}^W = H_{c_1}^W H_{c_2}^W \quad \mathbb{P} - a.s.$$

PROOF. This follows immediately from the following facts: $c^\perp$ depends linearly on $c$, the double layer potential is additive, the map $c \mapsto \sigma(c)$ is also additive and the map $c \mapsto U_1^{\lambda_1} \ldots U_{2g}^{\lambda_{2g}} x_1^{\nu_1} \ldots x_{p-1}^{\nu_{p-1}}$ is multiplicative. □

REMARK 3.12. Observe however that we do not claim to have a multiplicative version of our process. This would require an additive version of the white noise which does not exist. We are only concerned here with the *distribution* of the process.

We are now able to prove Theorem 3.2.

PROOF OF THEOREM 3.2. According to Proposition 2.32, we can approximate any finite family of cycles by piecewise geodesic cycles. The continuity properties of our families of variables (Theorem 2.72 and Proposition 3.4) show that it suffices to check the equality of distributions on piecewise geodesic cycles.

Any finite family of piecewise cycles can be realized as family of cycles in a graph $\Gamma$ and we know by Lemma 1.34 that it is enough to prove the equality of distributions for the fundamental system $(\ell_1, \ldots, \ell_{2g}, N_1, \ldots, N_p, \partial F_1, \ldots, \partial F_n)$, where $F_1, \ldots, F_n$ are the faces of the graph.

On one hand, $\ell_1^\perp = \ldots = \ell_{2g}^\perp = 0$, so that $H_{\ell_i}^W = U_i$ for all $i = 1, \ldots, 2g$. On the other hand, $N_1^\perp = \ldots = N_{p-1}^\perp = 0$ and $N_p^\perp = N_1 + \ldots + N_p$. Thus, $\sigma(N_p^\perp) = 1$ and $u_{N_p^\perp}$ is equal to zero so that $W_0(u_{N_p^\perp}) = 0$. This implies $H_{N_j}^W = x_j$ for all $j = 1, \ldots, p-1$ and also for $j = p$. Finally, Proposition 1.42 shows that $(H_{\partial F_1}^W, \ldots, H_{\partial F_n}^W)$ and $(\mathfrak{H}_{\partial F_1}, \ldots, \mathfrak{H}_{\partial F_n})$ have the same law. Since $(\mathfrak{H}_{\partial F_1}, \ldots, \mathfrak{H}_{\partial F_n})$ and $(\mathfrak{H}_{\ell_1}, \ldots, \mathfrak{H}_{\ell_{2g}})$ are independent under the Yang-Mills measure, this terminates the proof. □

## 3.2. Small scale structure of the Yang-Mills field

In the first part of this chapter, we have explained how the data of a white noise on $M$ and some finite-dimensional alea allows one to reconstruct the Yang-Mills measure when $G = U(1)$. We are now going to proceed backwards and extract a white noise from the Yang-Mills measure on a surface. Formally, this amounts to compute the curvature of a Yang-Mills random connection.

As usual, $(M, \sigma)$ is given, as well as elements $x_1, \ldots, x_p$ of $G$, associated with the components of $\partial M$ and $x = x_1 \ldots x_p$ or $x = 1$ if $M$ has no boundary.

In order to study the measure at a small scale, we construct on $M$ a sequence of partitions in the following way. Let $(\Gamma_n)_{n \geq 1}$ be a sequence of graphs on $M$ such that $\Gamma_n$ has exactly $n$ faces denoted by $F_{j,n}$, $j = 1, \ldots, n$. We assume that $\sigma(F_{j,n}) = \frac{\sigma(M)}{n}$ and also that the diameter of the faces decreases uniformly to 0, i.e. that for any metric on $M$, $\sup_j \text{diam}(F_{j,n}) \longrightarrow 0$. We fix an orientation of $M$ and assume that the boundaries of the faces are oriented with the usual convention. For each couple $(j, n)$ with $1 \leq j \leq n$, we denote the random variable $\mathfrak{H}_{\partial F_{j,n}}$ on $(\mathcal{M}(PM, G), \mathcal{I}, P_{(x_1,\ldots,x_p)})$ by $\mathfrak{H}_{j,n}$ and see it as a $\mathbb{C}$-valued random variable, by identifying $U(1)$ with $\{z \in \mathbb{C} : |z| = 1\}$.

For each $n \geq 1$, let $E_n$ denote the space of functions on $M$ constant on each face of $\Gamma_n$. Set $E_\infty = \cup_n E_n$. The assumption on the diameter of the faces $F_{j,n}$ imply that any continuous function on $M$ can be uniformly approximated by functions of $E_\infty$. Thus, $E_\infty$ is dense in $L^1(M, \sigma)$ and $L^2(M, \sigma)$.

We are now going to proceed as for the construction of the standard Wiener integral. Let us define a linear form $I_n$ on each $E_n$ as follows. Let $f_n$ be a function of $E_n$ and $f_{j,n}$ its value on $F_{j,n}$. Set

$$I_n(f_n) = \frac{1}{2i\pi} \sum_{j=1}^n f_{j,n}(\mathfrak{H}_{j,n} - 1) - i\pi \int_M f_n \, d\sigma.$$

REMARK 3.13. The imaginary term is meant to compensate the quadratic variation term which is going to appear. Indeed, let us derive informally the limit of $I_n(f_n)$ using the identity in distribution

$$\mathfrak{H}_{\partial D} = \exp 2i\pi \left( W_0(\mathbf{1}_D) + T\frac{\sigma(D)}{\sigma(M)} \right),$$

where $D$ is a good domain and $W_0, T$ are those used in the first part of this chapter. This relation can be written in a differential form as

$$\mathfrak{H}^{-1}d\mathfrak{H} = 2i\pi dW_0 + 2i\pi \frac{T}{\sigma(M)} d\sigma - 2\pi^2 d\sigma,$$

because $(dW_0)^2 = d\sigma$. This leads to

$$\lim I_n(f) = \frac{1}{2i\pi} \int_M f\, \mathfrak{H}^{-1} d\mathfrak{H} - i\pi \int_M f\, d\sigma = \int_M f\, dW_0 + \frac{T}{\sigma(M)} \int_M f\, d\sigma.$$

The next theorem is just a rigorous statement of this heuristic result.

THEOREM 3.14. *Let $f$ be a square-integrable function on $M$ and $(f_n)_{n\geq 0}$ a sequence of functions converging to $f$ in $L^2$ norm and such that $f_n \in E_n$. Then the sequence $(I_n(f_n))_{n\geq 1}$ converges in $L^2(\mathcal{M}(PM, G), \mathcal{I}, P_{(x_1,\ldots,x_p)})$ to a random variable $I(f)$ that does not depend on the choice of the sequence $(f_n)$. The law of this random variable can be described in the following way. Let $W_f^0$ be a centered Gaussian random variable with variance $\|f\|_{L_0^2}^2 = \|f - \frac{1}{\sigma(M)} \int_M f\, d\sigma\|_{L^2}^2$. Let $T$ be a $\mathcal{N}(0, \sigma(M))$ random variable conditioned to take its values in $\{t \in \mathbb{R} : e^{2i\pi t} = x\}$, independent of $W_f^0$. Then, the following identity holds in distribution:*

$$I(f) \stackrel{(d)}{=} W_f^0 + T \cdot \frac{1}{\sigma(M)} \int_M f\, d\sigma. \tag{3.1}$$

This proves in particular that the law of $I(f)$ does not depend on the choice of the orientation of $M$. We will comment this result at the end of this chapter.

PROOF. To prove this theorem, it is of course convenient to use the white noise realization of the Yang-Mills measure. Let $(\Omega, \mathcal{A}, \mathbb{P})$ be a probability space on which a pair $(W, T)$ is defined, consisting in a white noise $W$ and a random variable $T$ independent of $W$, whose law is that described in the theorem. We do not need the uniform variables $U_i$ introduced in the first part of this chapter, because we are only computing holonomies along loops that are homologous to zero. Set

$$Y_{j,n} = W(\mathbf{1}_{F_{j,n}}), \quad S_n = \sum_{j=1}^n Y_{j,n} = W(1), \quad X_{j,n} = Y_{j,n} - \frac{1}{n} S_n.$$

We know by Theorem 3.2 that the distribution of the sequence $(I_n(f_n))_{n\geq 1}$ can be represented on $(\Omega, \mathcal{A}, \mathbb{P})$ by the sequence

$$I_n^W = \frac{1}{2i\pi} \sum_{j=1}^n f_{j,n} \left( e^{2i\pi(X_{j,n} + \frac{T}{n})} - 1 \right) - i\pi \int_M f_n\, d\sigma, \quad n \geq 1.$$

It is equivalent to prove the theorem for this sequence. For this, we study the following Lagrange inequality:

$$\left| \sum_{j=1}^n f_{j,n} \left( e^{2i\pi(X_{j,n} + \frac{T}{n})} - 1 \right) - 2i\pi \sum_{j=1}^n f_{j,n} \left( X_{j,n} + \frac{T}{n} \right) + 2\pi^2 \sum_{j=1}^n f_{j,n} \left( X_{j,n} + \frac{T}{n} \right)^2 \right|$$

$$\leq 8\pi^3 \sum_{j=1}^{n} |f_{j,n}| \left| X_{j,n} + \frac{T}{n} \right|^3 \qquad (3.2)$$

We will use repeatedly the following two simple facts. The first one is the existence, for each positive integer $p$, of a constant $C_p$ such that

$$\mathbb{E}|X_{j,n}|^p \leq \frac{C_p}{n^{\frac{p}{2}}},$$

just because $X_{j,n}$ has variance $O(n^{-1})$. The second one is the finiteness of all moments of $T$.

During this proof, $C$ will denote a constant, i.e. a number that depends neither on $j$ nor on $n$. It may however denote different constants at different lines.

We begin by showing that the right hand side of (3.2) term converges to zero in $L^2(\Omega, \mathcal{A}, \mathbb{P})$. Observe that

$$\mathbb{E}\left| X_{j,n} + \frac{T}{n} \right|^6 \leq \frac{C}{n^3},$$

so that

$$\mathbb{E}\left| \sum_{j=1}^{n} |f_{j,n}| \left| X_{j,n} + \frac{T}{n} \right|^3 \right|^2 = \sum_{j,k=1}^{n} |f_{j,n}||f_{k,n}| \mathbb{E} \left| X_{j,n} + \frac{T}{n} \right|^3 \left| X_{k,n} + \frac{T}{n} \right|^3$$

$$\leq n^2 \sup_{j=1}^{n} \mathbb{E}\left| X_{j,n} + \frac{T}{n} \right|^6 \left( \sum_{j=1}^{n} \frac{|f_{j,n}|}{n} \right)^2$$

$$\leq \frac{C}{n} \| f_n \|_{L^1}^2 \longrightarrow 0.$$

Now let us look at the second order term of the left hand side of (3.2). Let $m_n = \frac{1}{\sigma(M)} \int_M f_n \, d\sigma$ denote the mean of $f_n$. We set $f_n^0 = f_n - m_n$. We will use several times the fact that $\sum_j f_{j,n}^0 = 0$. We have

$$\sum_{j=1}^{n} f_{j,n} \left( X_{j,n} + \frac{T}{n} \right)^2 = \sum_{j=1}^{n} f_{j,n}^0 \left( X_{j,n} + \frac{T}{n} \right)^2 + m_n \sum_{j=1}^{n} \left( X_{j,n} + \frac{T}{n} \right)^2. \qquad (3.3)$$

Let us study the first term of this decomposition.

$$\sum_{j=1}^{n} f_{j,n}^0 \left( X_{j,n} + \frac{T}{n} \right)^2 = \sum_{j=1}^{n} f_{j,n}^0 X_{j,n}^2 + 2 \sum_{j=1}^{n} \frac{f_{j,n}^0}{n} X_{j,n} T. \qquad (3.4)$$

The first term of the right hand side term can be written:

$$\sum_{j=1}^{n} f_{j,n}^0 X_{j,n}^2 = \sum_{j=1}^{n} f_{j,n}^0 (Y_{j,n} - \frac{1}{n} S_n)^2 = \sum_{j=1}^{n} f_{j,n}^0 Y_{j,n}^2 - 2 \sum_{j=1}^{n} \frac{f_{j,n}^0}{n} Y_{j,n} S_n.$$

On one hand,

$$\mathbb{E}\left| \sum_{j=1}^{n} f_{j,n}^0 Y_{j,n}^2 \right|^2 = \sum_{j=1}^{n} \frac{|f_{j,n}^0|^2}{n^2} \mathbb{E}|nY_{j,n}^2|^2 \leq \frac{C}{n} \| f_n^0 \|_{L^2}^2 \longrightarrow 0,$$

since $\mathbb{E}|nY_{j,n}|^2$ depends neither on $j$ nor on $n$. On the other hand,

$$\| Y_{j,n}S_n \|_{L^2}^2 = \mathbb{E}(Y_{j,n}^2 S_n^2) = (n-1)(\mathbb{E}Y_{j,n}^2)^2 + \mathbb{E}Y_{j,n}^4 \leq \frac{C}{n}$$

implies

$$\|\sum_{j=1}^n \frac{f_{j,n}^0}{n} Y_{j,n} S_n \|_{L^2} \leq \sum_{j=1}^n \frac{|f_{j,n}^0|}{n} \| Y_{j,n} S_n \|_{L^2} \leq \frac{C}{\sqrt{n}} \| f_n^0 \|_{L^1} \longrightarrow 0.$$

We have proved that the first term of the r.h.s. of (3.4) tends to 0. To study the second one, notice that

$$\| X_{j,n} T \|_{L^2}^2 = \mathbb{E} X_{j,n}^2 \mathbb{E} T^2 \leq \frac{C}{n},$$

so that

$$\|\sum_{j=1}^n \frac{f_{j,n}^0}{n} X_{j,n} T \|_{L^2} \leq \sum_{j=1}^n \frac{|f_{j,n}^0|}{n} \| X_{j,n} T \|_{L^2} \leq \frac{C}{\sqrt{n}} \| f_{j,n}^0 \|_{L^1} \xrightarrow[n\to\infty]{} 0.$$

We have proved that the zero-mean part of $f_n$ does not contribute to the second order term. Let us study the last term of (3.3).

$$m_n \sum_{j=1}^n \left( X_{j,n} + \frac{T}{n} \right)^2 = m_n \sum_{j=1}^n X_{j,n}^2 + 2\frac{m_n T}{n} \sum_{j=1}^n X_{j,n} + \frac{m_n}{n} T^2.$$

Using $\sum_j X_{j,n} = 0$ a.s. and $m_n/n \longrightarrow 0$, there remains

$$m_n \sum_{j=1}^n X_{j,n}^2 = m_n \sum_{j=1}^n Y_{j,n}^2 + \frac{m_n}{n} S_n^2 - 2\frac{m_n}{n} S_n.$$

Since $S_n = W(1)$ does not depend on $n$, the two last terms tend to zero. In order to determine the limit of the first one, we compute

$$\mathbb{E}\left|\sum_{j=1}^n Y_{j,n}^2 - \frac{\sigma(M)}{n}\right|^2 = \sum_{j=1}^n \mathbb{E}\left|Y_{j,n}^2 - \frac{\sigma(M)}{n}\right|^2 \leq \frac{C}{n^2}.$$

Thus,

$$m_n \sum_{j=1}^n Y_{j,n}^2 \xrightarrow{L^2} \sigma(M) \lim m_n = \int_M f \, d\sigma.$$

We are done with the second order term. We finish the proof by studying the first order one.

$$\sum_{j=1}^n f_{j,n}\left(X_{j,n} + \frac{T}{n}\right) = m_n T + \sum_{j=1}^n f_{j,n}^0 Y_{j,n}$$

$$= W(f_n^0) + m_n T \xrightarrow{L^2} W(f^0) + \frac{T}{\sigma(M)} \int_M f \, d\sigma.$$

We have proved that

$$I_n^W \xrightarrow{L^2} W(f^0) + T \cdot \frac{1}{\sigma(M)} \int_M f \, d\sigma.$$

This limit does not depend on the choice of the sequence $(f_n)$. Thus the sequence $(I_n(f_n))_{n\geq 1}$ converges also to a limit $I(f)$ that does not depend on the choice of $(f_n)$ and whose distribution is that claimed in the statement of the theorem. $\square$

This result confirms at a heuristic level our interpretation of $T$ as the total curvature of the random connection underlying the random holonomy process (see Section 1.9.3). Actually, we can now condition the process by the value of $T$, just by replacing $T$ with a deterministic number in Definition 3.1, and obtain in that way the distribution of the Yang-Mills measure living on a particular topological type of $U(1)$-bundle over $M$, or with a given total curvature if $M$ has a boundary.

# CHAPTER 4

# Small scale structure in the semi-simple case

Theorem 3.2 shows that it is possible to construct the Yang-Mills measure in a short and quite pleasant way when $G = U(1)$, using a white noise on $M$ as main ingredient. Is it possible to do something similar in general? The works of A. Sengupta and B. Driver [20, 36, 38] show how one can construct a random holonomy starting from a Lie algebra-valued white noise on $M$, but along a family of loops which is strongly dependent on a particular choice of coordinates on $M$ and lacks independence under area-preserving diffeomorphisms. The main result of this chapter hints at the fact that this is more than a technical complication.

Our idea is the following. Were it possible to extract a white noise from the random holonomy, it would be possible to find a lot of information by looking at the random holonomy at small scale, i.e. along very small loops. For instance, Theorem 3.14 basically says that when $G = U(1)$, almost all the information about the holonomy along homologically trivial loops is available at infinitesimally small scale. We prove that, when $G$ is semi-simple, there is no information at all available at infinitesimally small scale, at least when one looks at it in the same way that we did in the Abelian case.

The result presented in this chapter (see Theorem 4.1 and Corollary 4.2 for precise statements) can also be motivated from another point of view, which will also be adopted in Chapter 5. On the probability space of the Yang-Mills process, there is a natural way to localize the $\sigma$-field by setting, for all suitable subset $U$ of $M$,

$$\mathcal{I}_U = \sigma(\mathfrak{H}_{l_1,\ldots,l_n} : l_1,\ldots,l_n \in L_m U, m \in U, n \geq 1),$$

where $L_m U$ denotes of course the set of loops whose range is contained in $U$. A general question is to understand how these sub-$\sigma$-fields fit together into $\mathcal{I}$. In Chapter 5, we shall begin by proving the Markov property of the random field induced by the Yang-Mills process. This property says that, if $M$ is the union of two surfaces $M_1$ and $M_2$ with common boundary $N$, then $\mathcal{I}_{M_1}$ and $\mathcal{I}_{M_2}$ are independent conditional on $\mathcal{I}_N$.

Theorem 4.1 can be interpreted as a zero-one law for this Markov random field in the following sense. For all $\varepsilon > 0$ and all $U \subset M$, one can define a smaller $\sigma$-field by

$$\mathcal{I}_U^\varepsilon = \sigma(\mathfrak{H}_{l_1,\ldots,l_n} : l_1,\ldots,l_n \in L_m U, \sup_i \ell(l_i) < \varepsilon).$$

Then one can wonder what

$$\mathcal{I}_U^{0^+} = \bigcap_{\varepsilon > 0} \mathcal{I}_U^\varepsilon$$

looks like. Although we are not yet able to answer this question in this generality, Corollary 4.2 indicates that $\mathcal{I}_U^{0^+}$ might be trivial, in that it would contain only deterministic events.

## 4.1. Asymptotic independence: a zero-one law

**4.1.1. Statement of the result.** We consider as usual the surface $M$ endowed with a surface measure $\sigma$. Unless explicitly specified, we assume in this chapter that the group $G$ is compact, connected and semi-simple[1]. The simplest example of such a group is $G = SU(2)$. If $\partial M$ has $p$ components, we choose $p$ conjugacy classes $\mathfrak{x}_1, \ldots, \mathfrak{x}_p \in G/\mathrm{Ad}$ and consider the probability space $(\mathcal{M}(PM, G), \mathcal{I}, P_{(\mathfrak{x}_1, \ldots, \mathfrak{x}_p)})$ of the conditional Yang-Mills measure on $M$ with boundary conditions $\mathfrak{x}_1, \ldots, \mathfrak{x}_p$.

Let $L$ be a simple loop on $M$ which is the boundary of an open set $D$ diffeomorphic to a disk. For each $n \geq 0$, consider a graph $\Gamma_n$ on $D$ with exactly $4^n$ faces $F_{1,n}, \ldots, F_{4^n,n}$ such that $\sigma(F_{i,n}) = \frac{\sigma(D)}{4^n}$ for each $i$. Moreover, we assume that each face of $\Gamma_{n+1}$ is contained in a face of $\Gamma_n$. The fact that each face is cut into *four* pieces at each step is of course arbitrary. This situation is very similar to that described in Section 3.2, where we have computed the curvature of the random holonomy.

Were $G$ Abelian, we would have the equality of cycles $L = \partial F_{1,n} + \ldots + \partial F_{4^n,n}$ holding for each $n$, provided orientations are well chosen. This would imply $\mathfrak{H}_L = \mathfrak{H}_{\partial F_{1,n}} \ldots \mathfrak{H}_{\partial F_{4^n,n}}$ and for any function $f$ continuous on $G/\mathrm{Ad} = G$,

$$E[f(\mathfrak{H}_L)|\mathfrak{H}_{\partial F_{1,n}}, \ldots, \mathfrak{H}_{\partial F_{4^n,n}}] = f(\mathfrak{H}_L).$$

In that case, the knowledge of the holonomy along the 'small' loops $\partial F_{i,n}$ allows one to determine the holonomy along $L$. When $G$ is semi-simple, the situation is totally different.

For each $n \geq 0$, let $\mathcal{T}_n$ denote the $\sigma$-algebra generated by the variables $\mathfrak{H}_{\partial F_{i,n}}, 1 \leq i \leq 4^n$ and set $\mathcal{F}_n = \cup_{m \geq n} \mathcal{T}_m$. The sequence $\mathcal{F}_n$ is decreasing and we call $\mathcal{F}_\infty = \cap_{n \geq 0} \mathcal{F}_n$ its limit. The main result is the following:

THEOREM 4.1. *For any function $f$ continuous on $G/\mathrm{Ad}$, the following convergence holds:*

$$E[f(\mathfrak{H}_L)|\mathcal{F}_n] \xrightarrow[n \to \infty]{L^2} Ef(\mathfrak{H}_L) = E[f(\mathfrak{H}_L)|\mathcal{F}_\infty].$$

The main part of this result is that the last conditional expectation is a constant, which means that the holonomy along $L$ is independent of the $\sigma$-algebra $\mathcal{F}_\infty$. As explained in the introduction of this chapter, the following corollary is reminiscent of a zero-one law.

COROLLARY 4.2. *The $\sigma$-field $\mathcal{F}_\infty$ is trivial, in that it contains only events of probability 0 or 1.*

PROOF. Theorem 4.1 shows that the random variable $\mathfrak{H}_L$ is independent of $\mathcal{F}_\infty$. But we can also apply this theorem again by taking $\partial F_{1,1}$ as our new loop $L$ and we deduce that $\mathfrak{H}_{\partial F_{1,1}}$ is also independent of $\mathcal{F}_\infty$. In fact, we can do this with any variable $\mathfrak{H}_{\partial F_{i,n}}$ and we get the independence of $\mathcal{F}_\infty$ and $\sigma(\mathfrak{H}_{\partial F_{i,n}}, 1 \leq i \leq 4^n, n \geq 0)$. Since $\mathcal{F}_\infty$ is obviously contained in this last $\sigma$-field, $\mathcal{F}_\infty$ is independent of itself and the result is proved. □

---

[1]For a compact Lie group, semi-simplicity means that the center is a discrete, hence finite subgroup. Abelian and semi-simple compact Lie groups are the two main classes of compact Lie groups and any compact Lie group is isomorphic to the quotient by a finite central subgroup of a direct product $U(1)^k \times G$, where $G$ is semi-simple.

We will deduce Theorem 4.1 from the following weaker result:

THEOREM 4.3. *For any function $f$ continuous on $G/\mathrm{Ad}$, the following convergence holds:*
$$E[f(\mathfrak{H}_L)|\mathcal{T}_n] = E[f(\mathfrak{H}_L)|\mathfrak{H}_{\partial F_{1,n}},\ldots,\mathfrak{H}_{\partial F_{4^n,n}}] \xrightarrow[n\to\infty]{L^2} Ef(\mathfrak{H}_L).$$

It is really a weaker statement than Theorem 4.1, since the $\sigma$-algebra $\mathcal{T}_n$ is smaller than $\mathcal{F}_n$. In fact, we will use a nontrivial property of the Yang-Mills field to prove that 4.3 implies 4.1, namely its Markov property. This property will be stated, proved and studied in the next chapter, but we will not be able to use directly Theorem 5.1 in our proof. In few words, the Markov property says that, when the surface $M$ is cut into several parts by a loop or a family of loops, then the random holonomies along the loops contained in different parts of $M$ are independent, conditionnal on the holonomy along the loops that delimit the parts of $M$. Further details can be found in Section 5.1.

Let us explain informally how this property applies to our situation. The conditional expectation $E[f(\mathfrak{H}_L)|\mathcal{F}_n]$ can be written $E[f(\mathfrak{H}_L)|\mathcal{T}_n \vee \mathcal{F}_{n+1}]$. If we cut $M$ along the loops $\partial F_{i,n}, i = 1,\ldots,4^n$, the loop $L$ on one hand and the loops $\partial F_{i,m}, m \geq n+1$ on the other hand are contained in different pieces of $M$. Thus, $\mathfrak{H}_L$ is independent of $\mathcal{F}_{n+1}$ given $\mathcal{T}_n$, so that
$$E[f(\mathfrak{H}_L)|\mathcal{F}_n] = E[f(\mathfrak{H}_L)|\mathcal{T}_n \vee \mathcal{F}_{n+1}] = E[f(\mathfrak{H}_L)|\mathcal{T}_n],$$
and the convergence stated in 4.3 implies that stated in 4.1.

Unfortunately, Theorem 5.1 is stated only in the case of a surface cut out along several *disjoint* loops, and given $n$, the loops $\partial F_{1,n},\ldots,\partial F_{4^n,n}$ are not at all disjoint. Nevertheless, they do not cross each other transversally and can be approximated by disjoint loops. This will allow us to prove by hand the Markov property in this particular situation and this proof will provide us with a fairly explicit expression of the conditional expectations involved.

### 4.1.2. Markov property in a particular case.
Our goal is to prove the following result:

PROPOSITION 4.4. *For each $n \geq 0$, the random variable $\mathfrak{H}_L$ is independent of $\mathcal{F}_n$ conditional on $\mathcal{T}_n$.*

As explained above, this proposition implies that Theorems 4.3 and 4.1 are equivalent. As a byproduct, we will also get an expression of the conditional distribution of $\mathfrak{H}_L$ given $\mathcal{T}_n$.

PROOF. The integer $n \geq 0$ is fixed throughout the proof and we abbreviate $F_{i,n}$ in $F_i$ for $i = 1,\ldots,4^n$. Let $F'_1 = F_{i_1,n_1},\ldots,F'_m = F_{i_m,n_m}$ be $m$ arbitrary faces of any of the graphs finer than $\Gamma_n$. By definition, we want to show that, conditional on $\mathcal{T}_n$, $\mathfrak{H}_L$ is independent of the holonomies along the boundaries of these arbitrary faces.

For each $F_i$, consider a sequence $(L_{i,k})_{k\geq 0}$ of simple loops whose image is inside the interior of $F_i$ and such that $L_{i,k}$ converges in length, as $k$ tends to infinity, to $\partial F_i$. Now do the same for each $F'_j$, namely define a sequence $(L'_{j,l})_{l\geq 0}$ of simple loops contained in $F'_j$ converging to $\partial F'_j$ but in such a way that $L'_{j,l}$ is also in the interior of $L_{i,k}$ whenever $F'_j \subset F_i$ and $l \leq k$. This means that the $L'_{j,l}$ do not converge faster towards the $F_j'$ than the $L_{i,k}$ towards the $F_i$. Figure 1 below, where $L$ is the big square and $n = 1$, might help to get the point.

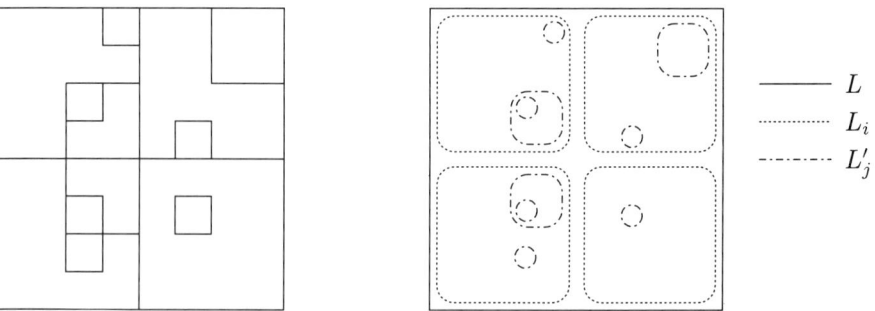

FIGURE 1. Approximation of the boundaries of the faces.

Theorem 2.72 shows that the following convergence holds in probability:

$$(\mathfrak{H}_{L_{1,k}}, \ldots, \mathfrak{H}_{L_{4^n,k}}, \mathfrak{H}_{L'_{1,l}}, \ldots, \mathfrak{H}_{L'_{m,l}}) \xrightarrow[k,l \to \infty]{P} (\mathfrak{H}_{\partial F_1}, \ldots, \mathfrak{H}_{\partial F_{4^n}}, \mathfrak{H}_{\partial F'_1}, \ldots, \mathfrak{H}_{\partial F'_m}).$$

Let $f$, $f_\mathcal{T}$ and $f_\mathcal{F}$ be continuous functions on $G/\mathrm{Ad}$, $(G/\mathrm{Ad})^{4^n}$ and $(G/\mathrm{Ad})^m$ respectively. The latter convergence implies the following one:

$$E[f(\mathfrak{H}_L) f_\mathcal{F}(\mathfrak{H}_{\partial F'_1}, \ldots, \mathfrak{H}_{\partial F'_m}) f_\mathcal{T}(\mathfrak{H}_{\partial F_1}, \ldots, \mathfrak{H}_{\partial F_{4^n}})] =$$
$$= \lim_{k,l \to \infty} E[f(\mathfrak{H}_L) f_\mathcal{F}(\mathfrak{H}_{L'_{1,l}}, \ldots, \mathfrak{H}_{L'_{m,l}}) f_\mathcal{T}(\mathfrak{H}_{L_{1,k}}, \ldots, \mathfrak{H}_{l_{4^n,k}})].$$

Let us fix $k$ and $l$, with $l \leq k$, and compute expectation on the right hand side. As long as $k$ and $l$ are fixed, we do not write the corresponding subscripts.

We construct a particular graph $\Gamma$ on $M$ such that $L, L_1, \ldots, L_{4^n}$ and $L'_1, \ldots, L'_m$ belong to $\Gamma^*$ and such that $\Gamma$ has only one face outside $D$. We fix a reference point $m_0$ on $M$. The support of $\Gamma$ contains the components $N_1, \ldots, N_p$ of $\partial M$ as well as $p$ paths $c_1, \ldots, c_p$ joining $m_0$ to them. It also contains $2g$ simple loops $a_1, \ldots, a_g, b_1, \ldots, b_g$ based at $m_0$ that represent a basis of the $H_1$ of a minimal closure of $M$. Finally, it contains a path $c$ which joins $m_0$ to $L(0)$. The boundary of the unique face of $\Gamma$ outside $D$ is $cL^{-1}c^{-1}c_p N_p c_p^{-1} \ldots c_1 N_1 c_1^{-1}[b_g, a_g] \ldots [b_1, a_1]$, where $[a, b]$ denotes the commutator $aba^{-1}b^{-1}$ of $a$ and $b$.

Inside $D$, the support of $\Gamma$ contains $4^n$ paths $d_1, \ldots, d_{4^n}$ joining $L(0)$ to the $L_i(0)$. These paths meet pairwise only at $L(0)$. It contains also paths joining the base point of each loop $L'_j$ to the base point of its parent in the natural tree structure induced by the inclusion on the set of loops $L, L_i, L'_j$. Figure 2 below should make our meaning clearer.

Still inside $D$, $\Gamma$ has a face with boundary $Ld_1 L_1^{-1} d_1^{-1} \ldots d_{4^n} L_{4^n}^{-1} d_{4^n}^{-1}$, whose area we will denote by $t$. When $k$ and $l$ tend to infinity, $t$ tends of course to zero. The other faces inside $D$ involve the loops $L'_j$ and we do not need to get into a precise description of what they are.

Let us compute $E[f(\mathfrak{H}_L) f_\mathcal{F}(\mathfrak{H}_{L'_1}, \ldots, \mathfrak{H}_{L'_m}) f_\mathcal{T}(\mathfrak{H}_{L_1}, \ldots, \mathfrak{H}_{L_{4^n}})]$ in $\Gamma$. Using the notation $[g]$ for the conjugacy class of an element $g$ of $G$ and the notation $\delta_{\mathfrak{x}}$ introduced in Section 1.5.3, we find

$$\frac{1}{Z(\mathfrak{x}_1, \ldots, \mathfrak{x}_p)} \int_{G^\Gamma} f([h_L]) f_\mathcal{F}([h_{L'_1}], \ldots, [h_{L'_m}]) f_\mathcal{T}([h_{L_1}], \ldots, [h_{L_{4^n}}])$$

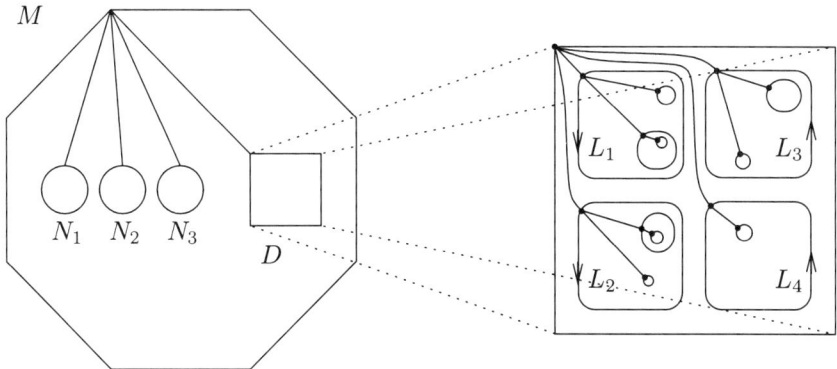

FIGURE 2. Aspect of $\Gamma$ outside and inside $D$.

$$P_1 \ldots P_{4^n} \, p_t(h_{d_{4^n}}^{-1} h_{L_{4^n}}^{-1} h_{d_{4^n}} \ldots h_{d_1}^{-1} h_{L_1}^{-1} h_{d_1} h_L)$$
$$p_{\sigma(D^c)}([h_{a_1}^{-1}, h_{b_1}^{-1}] \ldots [h_{a_g}^{-1}, h_{b_g}^{-1}] h_{c_1}^{-1} h_{N_1} h_{c_1} \ldots h_{c_p}^{-1} h_{N_p} h_{c_p} h_c^{-1} h_L^{-1} h_c)$$
$$d\nu_{x_1} \ldots d\nu_{x_p} d\delta_{(\mathfrak{x}_1,\ldots,\mathfrak{x}_p)}(x_1, \ldots, x_p) dg',$$

where $P_1, \ldots, P_n$ are the factors corresponding to the faces inside the interior of the $L_i$'s. Using the fact that, under $\nu_{x_1} \ldots \nu_{x_p} dg'$, the variables $h_L$, $h_{L_i}$, $h_{a_i}$, $h_{b_i}$, $h_{c_i}$, $h_c$ and $h_{d_i}$ are uniform and independent, we can rewrite this as

$$\frac{1}{Z(\mathfrak{x}_1,\ldots,\mathfrak{x}_p)} \int_{G^{2(g+4^n)+p+m+1}} f([g]) f_{\mathcal{F}}([g_1'], \ldots, [g_n']) f_{\mathcal{T}}([g_1], \ldots, [g_{4^n}])$$
$$P_1(g_1, \{g_j'\}) \ldots P_{4^n}(g_n, \{g_j'\}) \, dg_1' \ldots dg_m'$$
$$p_t(y_{4^n}^{-1} g_{4^n}^{-1} y_{4^n} \ldots y_1^{-1} g_1^{-1} y_1 g) \, dy_1 \ldots dy_{4^n} \, dg_1 \ldots dg_{4^n}$$
$$p_{\sigma(D^c)}([a_1, b_1] \ldots [a_g, b_g] x_1 \ldots x_p g^{-1}) \, da_1 \ldots da_g \, db_1 \ldots db_g$$
$$dg \, d\delta_{(\mathfrak{x}_1,\ldots,\mathfrak{x}_p)}(x_1, \ldots, x_p). \qquad (4.1)$$

We have used the right invariance of $dy_i$ to get rid of the term $h_c$. Beware the fact that $dg'$ refers to the edges inside the interior of $M$ in the first expression, while $g_1', \ldots, g_m'$ denote the holonomies along the $L_j'$'s in the second. Finally, recall that the letter $g$ in $[a_g, b_g]$ denotes the genus of $M$!

If we let $k$ tend to infinity, $t$ tends to zero and, according to Proposition 1.13, we can drop $g$ and replace it by $y_1^{-1} g_1 y_1 \ldots y_{4^n}^{-1} g_{4^n} y_{4^n}$. The functions $P_1, \ldots, P_{4^n}$ also depend on $k$, in fact they depend smoothly on the areas of the faces inside each loop $L_i$, provided these areas do not tend to zero, which is the case since $l$ remains fixed. So, they converge to some functions which we still denote by $P_1, \ldots, P_{4^n}$ and the property of main interest to us of these functions is that

$$\int_{G^m} P_i(g_i, \{g_j'\}) \, dg_1' \ldots dg_m' = p_{\frac{\sigma(D)}{4^n}}(g_i),$$

which is merely the invariance under subdivision of the discrete theory on the disk. At this point, we can give an expression of the conditional expectation of $f(\mathfrak{H}_L) f_{\mathcal{F}}(\mathfrak{H}_{L_1'}, \ldots, \mathfrak{H}_{L_m'})$ given $\mathfrak{H}_{\partial F_{1,n}} = \mathfrak{h}_1, \ldots, \mathfrak{H}_{\partial F_{4^n,n}} = \mathfrak{h}_{4^n}$ for some elements $\mathfrak{h}_1, \ldots, \mathfrak{h}_{4^n}$ in $G/\mathrm{Ad}$. We get:

$$E[f(\mathfrak{H}_L)f_{\mathcal{F}}(\mathfrak{H}_{L'_1},\ldots,\mathfrak{H}_{L'_m})|\mathfrak{H}_{\partial F_{1,n}} = \mathfrak{h}_1,\ldots,\mathfrak{H}_{\partial F_{4^n,n}} = \mathfrak{h}_{4^n}] =$$
$$p_{\frac{\sigma(D)}{4^n}}(\mathfrak{h}_1)^{-1}\ldots p_{\frac{\sigma(D)}{4^n}}(\mathfrak{h}_{4^n})^{-1}$$
$$\int_{G^m} f_{\mathcal{F}}(g'_1,\ldots,g'_m) P_1(\mathfrak{h}_1,\{g'_j\})\ldots P_{4^n}(\mathfrak{h}_{4^n},\{g'_j\})\, dg'_1\ldots dg'_m$$
$$\left(\int p_{\sigma(D^c)}([a_1,b_1]\ldots[a_g,b_g]x_1\ldots x_p k_{4^n}^{-1}\ldots k_1^{-1})\, da_1 db_1 \ldots da_g db_g\right.$$
$$\left. d\delta_{(\mathfrak{x}_1,\ldots,\mathfrak{x}_p)}(x_1,\ldots,x_p)\, d\delta_{(\mathfrak{h}_1,\ldots,\mathfrak{h}_{4^n})}(k_1,\ldots,k_{4^n})\right)^{-1}$$
$$\int f(k_1\ldots k_{4^n}) p_{\sigma(D^c)}([a_1,b_1]\ldots[a_g,b_g]x_1\ldots x_p k_{4^n}^{-1}\ldots k_1^{-1})\, da_1 db_1 \ldots da_g db_g$$
$$d\delta_{(\mathfrak{x}_1,\ldots,\mathfrak{x}_p)}(x_1,\ldots,x_p)\, d\delta_{(\mathfrak{h}_1,\ldots,\mathfrak{h}_{4^n})}(k_1,\ldots,k_{4^n}).$$

It appears that this is the product of the conditional expectations of $f(\mathfrak{H}_L)$ and $f'(\mathfrak{H}_{L'_1},\ldots,\mathfrak{H}_{L'_m})$, which are therefore independent given $\mathcal{T}_n$. Now, letting $l$ tend to infinity gives the desired conditional independence, namely that of $f(\mathfrak{H}_L)$ and $f_{\mathcal{F}}(\mathfrak{H}_{\partial F'_1},\ldots,\mathfrak{H}_{\partial F'_m})$ given $\mathcal{T}_n$, and finishes the proof. □

This proof gives us a very useful expression of the conditional expectation of $f(\mathfrak{H}_L)$ with respect to the $\sigma$-algebra $\mathcal{T}_n$.

PROPOSITION 4.5. *The following equality holds:*
$$E_{P_{(\mathfrak{x}_1,\ldots,\mathfrak{x}_p)}}[f(\mathfrak{H}_L)|\mathfrak{H}_{\partial F_{1,n}},\ldots,\mathfrak{H}_{\partial F_{4^n,n}}] =$$
$$\left(\int p_{\sigma(D^c)}([a_1,b_1]\ldots[a_g,b_g]x_1\ldots x_p k_{4^n}^{-1}\ldots k_1^{-1})\, da_1 db_1 \ldots da_g db_g\right.$$
$$\left. d\delta_{(\mathfrak{x}_1,\ldots,\mathfrak{x}_p)}(x_1,\ldots,x_p)\, d\delta_{(\mathfrak{H}_1,\ldots,\mathfrak{H}_{4^n})}(k_1,\ldots,k_{4^n})\right)^{-1}$$
$$\int f(k_1\ldots k_{4^n}) p_{\sigma(D^c)}([a_1,b_1]\ldots[a_g,b_g]x_1\ldots x_p k_{4^n}^{-1}\ldots k_1^{-1})\, da_1 db_1 \ldots da_g db_g$$
$$d\delta_{(\mathfrak{x}_1,\ldots,\mathfrak{x}_p)}(x_1,\ldots,x_p)\, d\delta_{(\mathfrak{H}_1,\ldots,\mathfrak{H}_{4^n})}(k_1,\ldots,k_{4^n}),$$
*where* $\mathfrak{H}_1 = \mathfrak{H}_{\partial F_{1,n}},\ldots,\mathfrak{H}_{4^n} = \mathfrak{H}_{\partial F_{4^n,n}}$.

In order to prove Theorem 4.3, we are going to study the limit of this expression as $n$ tends to infinity by using harmonic analysis on $G$. We give a very short account of the results that we will use. For a detailed presentation of the subject, see for example [16, 40].

**4.1.3. Characters of a compact Lie group.** A representation $(\rho, V)$ of $G$ is a smooth morphism of groups $\rho$ from $G$ into the linear group of some finite-dimensional complex linear space $V$. The *dimension* of $\rho$ is by definition that of $V$. Since $G$ is compact, we may always assume that $V$ is endowed with such a Hermitian scalar product that $\rho$ maps $G$ into the unitary group. A representation is *irreducible* if there are no subspaces of $V$ invariant by all $\rho(g)$ except the trivial ones, $V$ and $0$. Two representations $(\rho_1, V_1)$ and $(\rho_2, V_2)$ are *equivalent* if there exists a linear isomorphism $\phi : V_1 \longrightarrow V_2$ such that $\rho_2 \circ \phi = \phi \circ \rho_1$. The character of $\rho$ is the $\mathbb{C}$-valued function $\chi_\rho$ defined on $G$ by $\chi_\rho(g) = \text{tr}\,\rho(g)$. Two representations are equivalent if and only if they have the same character. Observe that $\chi_\rho$ is a central function on $G$, that is, $\chi_\rho(g)$ depends only on the conjugacy class of $g$, $\chi_\rho(1) = \dim \rho$, and, since $\rho(g)$ is unitary, $\chi_\rho(g^{-1}) = \overline{\chi_\rho(g)}$.

THEOREM 4.6 (Peter-Weyl theorem). *The set of characters of all equivalence classes of irreducible representations of $G$ is an orthonormal basis of the space of central square-integrable functions on $(G, dg)$. Moreover, the complex algebra generated by this set is dense in the set of continuous complex central functions on $G$ endowed with the uniform norm.*

According to this theorem, any continuous function on $G$ can be approximated by linear combinations of products of characters. In fact, a product of characters is always equal to a linear combination of characters of irreducible representations, so that the linear span of the characters is actually dense in $C(G)$.

We are going to prove Theorem 4.3 when $f$ is the character of an irreducible representation and the genreal case will follow by linearity and uniform approximation.

The orthogonality properties of the characters give rise to useful formulas. We will mainly use two of them.

PROPOSITION 4.7. *For any $x, y, a, b \in G$ and for any irreducible representation $\alpha$ of $G$,*

$$\int_G \chi_\alpha(zxz^{-1}y)\, dz = \frac{1}{\dim \alpha} \chi_\alpha(x) \chi_\alpha(y), \tag{4.2}$$

$$\int_{G^2} \chi_\alpha([a,b]x)\, da\, db = \frac{1}{(\dim \alpha)^2} \chi_\alpha(x). \tag{4.3}$$

We can rewrite (4.2) as

$$\int_G \chi_\alpha(xy)\, d\delta_{\mathfrak{x}}(x) = \frac{1}{\dim \alpha} \chi_\alpha(\mathfrak{x}) \chi_\alpha(y), \tag{4.4}$$

where $\mathfrak{x}$ is a conjugacy class in $G$ and $\delta_{\mathfrak{x}}$ is the $G$-invariant measure on $\mathfrak{x}$ introduced in 1.5.3.

Let us endow $G$ with its bi-invariant metric normalized to have total volume equal to 1. This metric gives rise to a Laplace operator on $G$. A remarkable property of the characters is that they are eigenfunctions for this operator. More precisely, for each irreducible representation $\alpha$, there exists a positive real number $c_\alpha$ such that

$$\Delta \chi_\alpha = -c_\alpha \chi_\alpha.$$

The classical theory of representations gives a full description of the irreducible representations – *irreps* – of $G$, in particular of their dimensions and the eigenvalues attached to their characters. We shall only need the fact that the irreps can be labeled by positive integers, $n \mapsto \beta_n$, in such a way that the dimension of $\beta_n$ grows not faster than a given power of $n$ and the eigenvalue $c_{\beta_n}$ grows not slower than another positive power of $n$.

A nice application of these properties is the computation of the character expansion of the heat kernel on $G$. Let us denote by $\widehat{G}$ the set of classes of irreducible representations of $G$.

PROPOSITION 4.8. *For all $t > 0$, the following equality holds uniformly and in $L^2(G, dg)$:*

$$p_t = \sum_{\beta \in \widehat{G}} \dim \beta \, e^{-\frac{c_\beta}{2} t} \chi_\beta.$$

PROOF. Using the relation $(\frac{1}{2}\Delta - \partial_t) p_t = 0$, one checks easily that the Fourier coefficients $\hat{p}_\beta(t) = (p_t, \chi_\beta)_{L^2}$ satisfy the differential equation $(\frac{1}{2} c_\beta + \partial_t) \hat{p}_\beta(t) = 0$. The value of $\hat{p}_\beta(0)$ is determined by the fact that $(p_t, \chi_\beta)_{L^2} \longrightarrow \chi_\beta(1)$ as $t$ tends to zero (see (1.7)).

The rough estimates on $\dim \beta$ and $c_\beta$ mentionned above imply the normal convergence of the Fourier series. □

**4.1.4. Expansion of the conditional expectation.** Let us consider the big expression obtained in Proposition 4.5. From now on, we fix an irreducible representation $\alpha$ of $G$ and take $f = \chi_\alpha$. Let us start by computing the numerator $N_n$ of the conditional expectation[2].

We begin by expanding the heat kernel $p_{\sigma(D^c)}$ using Proposition 4.8. We get a sum over $\beta$ of integrals of $\chi_\alpha$ of something times $\chi_\beta$ of something else. We integrate over the variables that appear only as arguments of $\chi_\beta$, first $a_1$ and $b_1$, and so on until $a_g$ and $b_g$, using (4.3). Each integration against $a_i$ and $b_i$ produces a factor $\frac{1}{(\dim \beta)^2}$. Then we integrate against $x_1, \ldots, x_p$ using (4.4). At each step, we get a factor $\chi_\beta(\mathfrak{x}_i) / \dim \beta$. At this stage, the arguments of $\chi_\alpha$ and $\chi_\beta$ under the integral are inverse of each other. Using the relation $\chi_\beta(g^{-1}) = \overline{\chi_\beta(g)}$, we find

$$N_n = \sum_{\beta \in \widehat{G}} (\dim \beta)^{1-2g-p} \, e^{-\frac{c_\beta}{2} \sigma(D^c)} \prod_{i=1}^{p} \chi_\beta(\mathfrak{x}_i) \cdot$$

$$\int_{G^n} \chi_\alpha \overline{\chi_\beta}(k_1 \ldots k_{4^n}) \, d\delta_{(\mathfrak{H}_1, \ldots, \mathfrak{H}_{4^n})}(k_1, \ldots, k_{4^n}).$$

Now we expand the product $\chi_\alpha \overline{\chi_\beta}$ as $\sum_{\gamma \in \widehat{G}} (\chi_\alpha \overline{\chi_\beta}, \chi_\gamma)_{L^2} \chi_\gamma$. Note that this is a finite sum corresponding to the decomposition of $\alpha \otimes \beta^\vee$ into irreps, where $\beta^\vee$ is the contragredient representation of $\beta$. We get a sum over $\beta$ and $\gamma$ but only $\chi_\gamma(k_1 \ldots k_{4^n})$ remains under the integral. We can use again (4.4) to integrate against $k_1, \ldots, k_{4^n}$. Notice that the factor $\frac{1}{\dim \gamma}$ produced by the last integration cancels out with the remaining $\chi_\gamma(1)$. The final result is

$$N_n = \sum_{\beta, \gamma \in \widehat{G}} (\dim \beta)^{1-2g-p} (\dim \gamma)^{-(4^n-1)} \, e^{-\frac{c_\beta}{2} \sigma(D^c)} (\chi_\alpha, \chi_\beta \chi_\gamma)_{L^2}$$

$$\prod_{i=1}^{p} \chi_\beta(\mathfrak{x}_i) \prod_{i=1}^{4^n} \chi_\gamma(\mathfrak{H}_{\partial F_{i,n}}). \quad (4.5)$$

---

[2]The computations involving characters presented in this section and the next one are very close to those done by Witten in [44], when he expands explicitly partition functions in order to compute the symplectic volume of the moduli space of flat connections.

If $\alpha$ is the trivial representation, $\chi_\alpha$ is identically equal to 1 and we get the following expression for the denominator $D_n$ of Proposition 4.5:

$$D_n = \sum_{\beta \in \widehat{G}} (\dim \beta)^{2-2g-p-4^n} e^{-\frac{c_\beta}{2}\sigma(D^c)} \prod_{i=1}^{p} \chi_\beta(\mathfrak{x}_i) \prod_{i=1}^{4^n} \overline{\chi_\beta}(\mathfrak{H}_{\partial F_{i,n}}). \quad (4.6)$$

In order to prove the convergence, we need to understand the asymptotic behaviour of

$$\frac{1}{(\dim \gamma)^{4^n-1}} \prod_{i=1}^{4^n} \chi_\gamma(\mathfrak{H}_{\partial F_i}).$$

The fact that $\overline{\chi_\beta}$ appears in (4.6) instead of $\chi_\beta$ is not a problem, since $\overline{\chi_\beta}$ is also the character of an irreducible representation, the contragredient of $\beta$. Now the main convergence result is the following.

PROPOSITION 4.9. *For each $\beta \in \widehat{G}$, the following convergence holds:*

$$\frac{1}{(\dim \beta)^{4^n-1}} \prod_{i=1}^{4^n} \chi_\beta(\mathfrak{H}_{\partial F_{i,n}}) \xrightarrow[n \to \infty]{L^2} \dim \beta \; e^{-\frac{c_\beta}{2}\sigma(D)}.$$

This result will be proved in the next section. Let us show that it implies Theorem 4.1.

PROOF OF THEOREM 4.1. The presence of the exponential term in the sum allows us to swap summations and limits.

By combining Proposition 4.9 and (4.5), we find the following $L^2$-limit for the numerator:

$$\sum_{\beta,\gamma \in \widehat{G}} (\dim \beta)^{1-2g-p} e^{-\frac{c_\beta}{2}\sigma(D^c)} \dim \gamma \; e^{-\frac{c_\gamma}{2}\sigma(D)} (\chi_\alpha, \chi_\beta \chi_\gamma)_{L^2} \prod_{i=1}^{p} \chi_\beta(\mathfrak{x}_p).$$

By the same kind of arguments that we have used to derive (4.5), it is easy to check that this expression is equal to the following:

$$\int_{G^{p+2g+1}} \chi_\alpha(g) p_{\sigma(D)}(g^{-1}) p_{\sigma(D^c)}([a_1,b_1]\ldots[a_g,b_g]x_1\ldots x_p g^{-1})$$
$$d\delta_{(\mathfrak{x}_1,\ldots,\mathfrak{x}_p)}(x_1,\ldots,x_p) \, da_1 db_1 \ldots da_g db_g \, dg. \quad (4.7)$$

For the denominator, we find similarly the following $L^2$-limit:

$$\sum_{\beta \in \widehat{G}} (\dim \beta)^{2-2g-p} e^{-\frac{c_\beta}{2}(\sigma(D^c)+\sigma(D))} \prod_{i=1}^{p} \chi_\beta(\mathfrak{x}_p),$$

which is equal to

$$\int_{G^{p+2g}} p_{\sigma(M)}([a_1,b_1]\ldots[a_g,b_g]x_1\ldots x_p) \, d\delta_{(\mathfrak{x}_1,\ldots,\mathfrak{x}_p)}(x_1,\ldots,x_p) \, da_1 db_1 \ldots da_g db_g \quad (4.8)$$

and this is nothing but $Z_M(\mathfrak{x}_1,\ldots,\mathfrak{x}_p)$.

Using finally the fact that $p_{\sigma(D)}(g^{-1}) = p_{\sigma(D)}(g)$, we see that the quotient of (4.7) by (4.8) is equal to $E_{P_{(\mathfrak{x}_1,\ldots,\mathfrak{x}_p)}}[\chi_\alpha(\mathfrak{H}_L)]$. This proves the theorem, up to Proposition 4.9, which shall be proved in the remainder of the chapter. □

## 4.2. Asymptotic independence on the plane

In order to prove Proposition 4.9, we begin with the following result, which can be seen as a reformulation of the property of asymptotic independence when the manifold $M$ is the plane $\mathbb{R}^2$.

PROPOSITION 4.10. *Let $(B_t^n)_{t \in \mathbb{R}_+, n \geq 0}$ be a sequence of independent Brownian motions on $G$. For all irreducible representation $\beta$ of $G$ and all positive real number $T$, the following convergence holds in probability:*

$$\frac{1}{(\dim \beta)^{n-1}} \prod_{i=1}^{n} \chi_\beta(B_{\frac{T}{n}}^i) \xrightarrow[n \to \infty]{\mathrm{P}} \dim \beta \, e^{-\frac{c_\beta}{2} T}.$$

This is really our central result. It is the place where the fact that $G$ is semi-simple will be used, in the following way.

PROPOSITION 4.11. *Let $G$ be a semi-simple compact Lie group. Let $f$ be a central function on $G$ differentiable at $1$. Then*

$$d_1 f = 0.$$

PROOF. Since $f$ is a central function, $d_1 f$ is a linear form on $\mathfrak{g}$ invariant by adjunction and $\mathfrak{h} = \ker d_1 f$ is an ideal of $\mathfrak{g}$, that is, a linear subspace such that $[\mathfrak{g}, \mathfrak{h}] \subset \mathfrak{h}$. Recall that $\mathfrak{g}$ is endowed with an invariant scalar product and set $\mathfrak{t} = \mathfrak{h}^\perp$. Then $\mathfrak{t}$ is also an ideal as the orthogonal of an ideal and it has at most dimension 1, so that it is commutative and contained in the center of $\mathfrak{g}$. Since $G$ is semi-simple, this center is trivial and $\mathfrak{h} = \mathfrak{g}$. $\square$

REMARK 4.12. Semi-simple compact Lie groups have very small centers and hence large conjugacy classes. For example, the exponential mapping $\mathfrak{su}(2) \longrightarrow SU(2)$ realizes a diffeomorphism between a neighbourhood of the identity in $SU(2)$ and the unit ball in $\mathbb{R}^3$ in such a way that the conjugacy classes correspond exactly to the spheres centered at the origin. The result for $SU(2)$ follows immediately from this description.

The following lemma contains the main part of the analysis required to prove Proposition 4.10.

LEMMA 4.13. *Let $(B_t)_{t \geq 0}$ be a Brownian motion on $G$ and $\alpha$ an irreducible representation of $G$. There exists a constant $C$ such that a decomposition of the following form holds:*

$$\frac{\chi_\alpha(B_t)}{\dim \alpha} = 1 - \frac{c_\alpha t}{2} + Y_t + Z_t$$

*with $E|Y_t|^2 \leq Ct^4$, $EZ_t = 0$ and $E|Z_t|^2 \leq Ct^2$.*

PROOF. We represent $(B_t)$ as a solution of a Stratonovich stochastic differential equation [**29**]. Recall that the data of a bi-invariant metric on $G$ is equivalent to that of a scalar product on the Lie algebra of $G$, invariant by adjunction. Let $(X_1, \ldots, X_{\dim G})$ be a basis of $\mathfrak{g}$ orthonormal for this scalar product. Each $X_i$ is seen as a left-invariant vector field on $G$. Let $W^1, \ldots, W^{\dim G}$ be independent real Brownian motions. Then the Brownian motion on $G$ satisfies:

$$\begin{cases} dB_t = \sum_{i=1}^{\dim G} X_i \circ dW_t^i \\ B_0 = 1. \end{cases} \quad (4.9)$$

The meaning of this notation is that, for any function $f$ of class $C^2$ on $G$,

$$f(B_t) = f(1) + \sum_{i=1}^{\dim G} \int_0^t X_i f(B_s)\, dW_s^i + \frac{1}{2} \int_0^t \Delta f(B_s)\, ds.$$

We apply this relation to $f = \chi_\alpha$. Using $\chi_\alpha(1) = \dim \alpha$ and $\Delta \chi_\alpha = -c_\alpha \chi_\alpha$, it becomes:

$$\frac{\chi_\alpha(B_t)}{\dim \alpha} = 1 - \frac{c_\alpha t}{2} + \frac{1}{\dim \alpha} \sum_{i=1}^{\dim G} \int_0^t X_i \chi_\alpha(B_s)\, dW_s^i$$

$$- \frac{c_\alpha}{2 \dim \alpha} \int_0^t (\chi_\alpha(B_s) - \dim \alpha)\, ds.$$

Set

$$Y_t = -\frac{c_\alpha}{2 \dim \alpha} \int_0^t (\chi_\alpha(B_s) - \dim \alpha)\, ds,$$

$$Z_t = \frac{1}{\dim \alpha} \sum_{i=1}^{\dim G} \int_0^t X_i \chi_\alpha(B_s)\, dW_s^i.$$

We keep the notation $\rho(g)$ for the bi-invariant distance $d(1, g)$ when $g \in G$. This $\rho$ has nothing to do with a representation of $G$! According to Proposition 4.11, $d_1 \chi_\alpha = 0$. This implies that $|\chi_\alpha(g) - \dim \alpha| = O(\rho(g)^2)$ in a neighbourhood of 1. By using Lemma 1.31, we get

$$\begin{aligned}
E|Y_t|^2 &= CE \left| \int_0^t (\chi_\alpha(B_s) - \dim \alpha)\, ds \right|^2 \\
&\leq Ct\, E \int_0^t |\chi_\alpha(B_s) - \dim \alpha|^2\, ds \\
&\leq Ct\, E \int_0^t \rho(B_s)^4\, ds \\
&\leq Ct^4. \qquad (4.10)
\end{aligned}$$

For each $i = 1, \ldots, \dim G$, the function $X_i \chi_\alpha$ is smooth and $X_i \chi_\alpha(1) = 0$. Thus, $|X_i \chi_\alpha(g)| = O(\rho(g))$ in a neighbourhood of 1. This gives

$$\begin{aligned}
E|Z_t|^2 &= CE \left| \sum_{i=1}^{\dim G} \int_0^t X_i \chi_\alpha(B_s)\, dW_s^i \right|^2 \\
&= C \sum_{i=1}^{\dim G} E \int_0^t |X_i \chi_\alpha(B_s)|^2\, ds \\
&\leq C \dim G \int_0^t E\rho(B_s)^2\, ds \\
&\leq Ct^2. \qquad (4.11)
\end{aligned}$$

This finishes the proof. $\square$

PROOF OF PROPOSITION 4.10. Let us consider a sequence $(B_t^n)$ defined on a probability space $(\Omega, \mathcal{F}, P)$. We are interested in the product

$$X_n = \frac{1}{(\dim \beta)^n} \prod_{i=1}^n \chi_\beta(B_{\frac{T}{n}}^i) = \prod_{i=1}^n (1 - \frac{c_\beta T}{2n} + Z_{T/n}^i + Y_{T/n}^i),$$

where the random variables with different exponents are independent, and we would like to take the logarithm of this product. This requires some precaution. Set

$$\Omega_n = \left\{ \left| Z_{T/n}^i + Y_{T/n}^i \right| < \frac{1}{3} \ \forall i = 1 \ldots n \right\}.$$

A simple Chebyshev inequality gives

$$P\left( |Z_t + Y_t| < \frac{1}{3} \right) \geq 1 - 9E|Z_t + Y_t|^2 \geq 1 - Ct^2$$

and implies

$$P(\Omega_n) \geq \left(1 - \frac{CT^2}{n^2}\right)^n \xrightarrow[n\to\infty]{} 1.$$

We do not change any convergence in distribution on $\Omega$ if we replace $X_n$ by 1 outside $\Omega_n$. So we set

$$\widetilde{X}_n = X_n \mathbf{1}_{\Omega_n} + \mathbf{1}_{\Omega_n^c}.$$

Then $\operatorname{Log} \widetilde{X}_n$ is well defined, if Log denotes the usual determination of the complex logarithm on $\mathbb{C} - \mathbb{R}^-$. In fact, we have more than that. If $n$ is such that $\frac{c_\beta T}{2n}$ is smaller than $\frac{1}{6}$, then each factor of $\widetilde{X}_n$ is of the form $(1-z)$ with $|z| < \frac{1}{2}$. For such a $z$, we have $|\operatorname{Log}(1-z) + z| \leq |z|^2$. Thus, the equality

$$\operatorname{Log}(\widetilde{X}_n) = \mathbf{1}_{\Omega_n} \sum_{i=1}^n \operatorname{Log}(1 - \frac{c_\beta T}{2n} + Z_{T/n}^i + Y_{T/n}^i)$$

implies

$$\left| \operatorname{Log} \widetilde{X}_n + \sum_{i=1}^n \left( \frac{c_\beta T}{2n} - Z_{T/n}^i - Y_{T/n}^i \right) \right| \leq \mathbf{1}_{\Omega_n} \sum_{i=1}^n \left| \frac{c_\beta T}{2n} - Z_{T/n}^i - Y_{T/n}^i \right|^2 +$$

$$\mathbf{1}_{\Omega_n^c} \left| \sum_{i=1}^n \frac{c_\beta T}{2n} - Z_{T/n}^i - Y_{T/n}^i \right|. \quad (4.12)$$

The last term tends to 0 in probability because $P(\Omega_n^c)$ tends to 0. Using (4.10) and (4.11), we find

$$E \left| \frac{c_\beta T}{2n} - Z_{T/n}^i - Y_{T/n}^i \right|^2 \leq \frac{CT^2}{n^2},$$

so that the first term of the right hand side of (4.12) tends to 0 in $L^1$ norm.

Now it follows easily from Lemma 4.13 that $\sum_i Y_{T/n}^i$ and $\sum_i Z_{T/n}^i$ tend to zero, in $L^1$ and $L^2$ respectively.

Finally, $\operatorname{Log} \widetilde{X}_n$ converges in probability to $-\frac{c_\beta T}{2}$ and the result follows. $\square$

There remains to prove that Proposition 4.10 implies Proposition 4.9.

PROOF OF PROPOSITION 4.9. Set $T = \sigma(D)$. Let $F$ be a bounded continuous function on $\mathbb{R}$. Let us compute

$$E_{P_{(\mathfrak{x}_1,\ldots,\mathfrak{x}_p)}}\left[F\left(\frac{1}{(\dim \beta)^{4^n-1}}\prod_{i=1}^{4^n}\chi_\beta(\mathfrak{H}_{\partial F_i})\right)\right]. \tag{4.13}$$

By taking the limit of (4.1) as $t$ tends to zero, one checks that it is equal to

$$\frac{1}{Z(\mathfrak{x}_1,\ldots,\mathfrak{x}_p)}\int F\left(\frac{1}{(\dim \beta)^{4^n-1}}\prod_{i=1}^{4^n}\chi_\beta(g_i)\right)\prod_{i=1}^{4^n}p_{\frac{\sigma(D)}{4^n}}(g_i)$$
$$p_{\sigma(D^c)}([a_g,b_g]\ldots[a_1,b_1]x_1\ldots x_p y_{4^n}^{-1}g_{4^n}^{-1}y_{4^n}\ldots y_1^{-1}g_1^{-1}y_1)$$
$$da_1 db_1 \ldots da_g db_g\, dy_1 dg_1 \ldots dy_{4^n} dg_{4^n}\, d\delta_{(\mathfrak{x}_1,\ldots,\mathfrak{x}_p)}(x_1,\ldots,x_p).$$

By expanding the heat kernel $p_{\sigma(D^c)}$ and integrating against all variables except $g_1,\ldots,g_{4^n}$, we find

$$\frac{1}{Z(\mathfrak{x}_1,\ldots,\mathfrak{x}_p)}\sum_{\gamma\in\widehat{G}}(\dim \gamma)^{1-2g-p}e^{-\frac{c_\gamma}{2}\sigma(D^c)}\prod_{i=1}^p\chi_\gamma(\mathfrak{x}_i)$$
$$\int_{G^n}F\left(\frac{1}{(\dim \beta)^{4^n-1}}\prod_{i=1}^{4^n}\chi_\beta(g_i)\right)\left[\frac{1}{(\dim \gamma)^{4^n-1}}\prod_{i=1}^{4^n}\chi_\gamma(g_i)\right]\prod_{i=1}^{4^n}p_{\frac{\sigma(D)}{4^n}}(g_i)\,dg_1\ldots dg_{4^n}.$$

Proposition 4.10 says exactly that this last integral converges to

$$F(\dim \beta\, e^{-\frac{c_\beta}{2}\sigma(D)})\dim \gamma\, e^{-\frac{c_\gamma}{2}\sigma(D)}.$$

Thus, as $n$ tends to infinity, (4.13) tends to

$$\frac{1}{Z(\mathfrak{x}_1,\ldots,\mathfrak{x}_p)}\sum_{\gamma\in\widehat{G}}(\dim \gamma)^{2-2g-p}e^{-\frac{c_\gamma}{2}\sigma(M)}\prod_{i=1}^p\chi_\gamma(\mathfrak{x}_i)\,F(\dim \beta\, e^{-\frac{c_\beta}{2}\sigma(D)})$$

and this is precisely equal to

$$F(\dim \beta\, e^{-\frac{c_\beta}{2}\sigma(D)})$$

by the observation made when we were computing (4.8).

This proves that the claimed convergence holds in distribution. Since the limit is deterministic and the sequence of random variables we consider is uniformly bounded by $\dim \beta$, the convergence also holds in quadratic mean. □

CHAPTER 5

# Surgery of the Yang-Mills measure

Our purpose in this chapter is to study how the Yang-Mills measure transforms under the surgery of surfaces. We study two basic operations, namely gluing two surfaces along a circle and gluing together two components of the boundary of a single surface. In both cases, we show that some additional information is needed in general to deduce the Yang-Mills measure on the resulting surface from that on the initial one(s). We characterize this information under the form of a $\sigma$-field independent of the initial data.

For the first operation, a special property of the measure arises, namely its Markov property. We have been led to use already twice some special cases of this property, in Section 2.9 and in Chapter 4. It had already been studied, for example by A. Sengupta and C. Becker in the discrete setting, in [12], and for a class of measures containing the discrete Yang-Mills measure by S. Albeverio et al. in [2].

We finish this chapter with a study of the conditional partition functions and we show in particular that they satisfy nice algebraic relations which are reminiscent of those appearing in topological quantum field theories.

## 5.1. Markov property of the Yang-Mills field

Consider a surface $M$ which is cut into two pieces $M_1$ and $M_2$. Our aim in the first part of this chapter is to understand the relationship between the Yang-Mills measures on $M_1$, $M_2$ and $M$. We have already met a question of this kind when we constructed the random holonomy on a surface with boundary starting from that on a minimal closure of this surface and this has led us to prove in a particular case the Markov property which is the object of Theorem 5.1.

Let $M_1$ and $M_2$ be two oriented surfaces with boundary such that $\partial M_1$ and $\partial M_2$ have each at least $p$ boundary components, where $p > 0$ is arbitrary. Pick $p$ components $N_1, \ldots, N_p$ of $\partial M_1$ and $p$ others $N'_1, \ldots, N'_p$ of $\partial M_2$. For each $i = 1, \ldots, p$, consider an orientation-reversing diffeomorphism $\psi_i : N_i \longrightarrow N'_i$ and call $M$ the surface obtained by gluing $M_1$ and $M_2$ along $\psi_1, \ldots, \psi_p$. Denote by $L_1, \ldots, L_p$ $p$ loops on $M$ whose images are the components of the common boundary of $M_1$ and $M_2$, oriented as boundary components of $M_1$.

Recall that the Yang-Mills measures on $M$, $M_1$ and $M_2$, denoted respectively by $P^M$, $P^{M_1}$ and $P^{M_2}$, are measures on the spaces $\mathcal{M}(PM, G)$, $\mathcal{M}(PM_1, G)$ and $\mathcal{M}(PM_2, G)$ endowed with the $\sigma$-algebras $\mathcal{I}$, $\mathcal{I}_1$ and $\mathcal{I}_2$ generated by the variables of the form $\mathfrak{H}_{l_1,\ldots,l_n}$, where $l_1, \ldots, l_n$ belong respectively to $LM$, $LM_1$ and $LM_2$.

We identify the $\sigma$-fields $\mathcal{I}_1$ and $\mathcal{I}_2$ with the sub-$\sigma$-fields $\sigma(\mathfrak{H}_{l_1,\ldots,l_n}, l_k \in LM_i)$, $i = 1, 2$ of $\mathcal{I}$. They coincide with the sub-$\sigma$-fields $\mathcal{I}_{M_1}$ and $\mathcal{I}_{M_2}$ defined in the introduction of Chapter 4.

THEOREM 5.1. *The $\sigma$-fields $\mathcal{I}_1$ and $\mathcal{I}_2$ are independent on $\mathcal{M}(PM, G)$ under $P^M$, given the random variable $(\mathfrak{H}_{L_1}, \ldots, \mathfrak{H}_{L_p})$. Moreover, let $f_1$ and $f_2$ be two measurable functions on $(\mathcal{M}(PM_1, G), \mathcal{I}_1)$ and $(\mathcal{M}(PM_2, G), \mathcal{I}_2)$ respectively. Then the product $f_1 f_2$ can be seen as an $\mathcal{I}$-measurable function on $\mathcal{M}(PM, G)$ and for all $\mathfrak{x}_1, \ldots, \mathfrak{x}_p \in G/\mathrm{Ad}$, the following equality holds:*

$$P^M_{(\mathfrak{x}_1, \ldots, \mathfrak{x}_p)}(f_1 f_2) = P^{M_1}_{(\mathfrak{x}_1, \ldots, \mathfrak{x}_p)}(f_1) P^{M_2}_{(\mathfrak{x}_1^{-1}, \ldots, \mathfrak{x}_p^{-1})}(f_2). \tag{5.1}$$

*Finally, these properties remain true if we condition further the Yang-Mills measures with respect to the random holonomy along some other boundary components or some interior loops on $M_1$ or $M_2$.*

This theorem says two things. It says that the random holonomy on $M_1$ is independent of that on $M_2$ given the holonomy along the common boundary of $M_1$ and $M_2$ and it says also that the restrictions to $\mathcal{I}_1$ and $\mathcal{I}_2$ of the measure $P^M_{(\mathfrak{x}_1, \ldots, \mathfrak{x}_p)}$ are equal, via the identification of $\sigma$-fields mentioned above, to the Yang-Mills measures on $M_1$ and $M_2$. Notice at this occasion that the inverse of a conjugacy class is still a conjugacy class, so that something like $\mathfrak{x}_1^{-1}$ is well-defined.

We prove first a discrete result, which is due, in a slightly different form, to C. Becker and A. Sengupta [12].

PROPOSITION 5.2. *Let $\Gamma$ be a graph on $M$ such that $L_1, \ldots, L_p$ belong to $\Gamma^*$. Let $\Gamma_1$ and $\Gamma_2$ be the graphs on $M_1$ and $M_2$ induced by $\Gamma$. Let $f$ be a continuous function on $G^{\Gamma_1} \times G^{\Gamma_2}$ invariant under both the gauge transformations on $\Gamma_1$ and $\Gamma_2$. Then $f$ gives rise to a gauge-invariant function on $G^\Gamma$, still denoted by $f$ and, for all $x \in G^p$,*

$$\int_{G^{\Gamma_1} \times G^{\Gamma_2}} f \, dP^{\Gamma_1}_x dP^{\Gamma_2}_{x^{-1}} = \int_{G^\Gamma} f \, dP^\Gamma_x.$$

*Moreover, this results remains true if one puts conditions on the holonomy along other components of the boundaries of $M_1$ or $M_2$ or along interior loops to $M_1$ and $M_2$.*

PROOF. By the Stone-Weierstrass theorem, it is enough to prove this proposition when $f$ is a product $f_1 f_2$ of two gauge-invariant functions on $G^{\Gamma_1}$ and $G^{\Gamma_2}$ respectively. In order to shorten the expressions, we abbreviate $d\nu_{x_1} \ldots d\nu_{x_p}$ into $d\nu_x$ if $x = (x_1, \ldots, x_p)$. We prove the proposition with further conditions on boundary components or interior loops on $M_1$ or $M_2$ represented by two collections $y_1$ and $y_2$ of elements of $G$. So, if $f = f_1 f_2$,

$$\int_{G^{\Gamma_1} \times G^{\Gamma_2}} f_1 f_2 \, dP^{\Gamma_1}_{(y_1, x)} dP^{\Gamma_2}_{(x^{-1}, y_2)} = \int_{G^{\Gamma_1}} f_1 \, dP^{\Gamma_1}_{(y_1, x)} \int_{G^{\Gamma_2}} f_2 \, dP^{\Gamma_2}_{(x^{-1}, y_2)}$$
$$= \frac{1}{Z_{M_1}(y_1, x)} \int_{G^{\Gamma_1}} f_1 D^{\Gamma_1} \, d\nu_{y_1} d\nu_x dg^{\Gamma'_1} \frac{1}{Z_{M_2}(x^{-1}, y_2)} \int_{G^{\Gamma_2}} f_2 D^{\Gamma_2} \, d\nu_{y_2} d\nu_{x^{-1}} dg^{\Gamma'_2}. \tag{5.2}$$

As usual, $dg^{\Gamma'_1}$ denotes the product measure over the set of edges of $\Gamma_1$ over which we do not put any condition. The two integrals in the last expression are very similar so it is enough to study the first one. Denote by $\Gamma_{1,\partial}$ the graph on $\partial M_1$ induced by $\Gamma_1$. The measure $d\nu_x$ concerns only the variables associated with the edges of $\Gamma_{1,\partial}$. More precisely,

$$\int_{G^{\Gamma_1}} f_1 D^{\Gamma_1} \, d\nu_{y_1} d\nu_x dg^{\Gamma'_1} = \int_{G^{\Gamma_{1,\partial}}} \left( \int_{G^{\Gamma_1 \setminus \Gamma_{1,\partial}}} f_1 D^{\Gamma_1} \, d\nu_{y_1} dg^{\Gamma'_1} \right) d\nu_x.$$

The function $I_1 = \int_{G^{\Gamma_1 \setminus \Gamma_{1,\partial}}} f_1 D^{\Gamma_1} \, d\nu_{y_1} dg^{\Gamma'_1}$ is a function on $G^{\Gamma_{\partial,1}}$ and we claim that it can be expressed as a function of $\mathfrak{H}_{L_1}, \ldots, \mathfrak{H}_{L_p}$.

This comes from the fact that $f_1 D^{\Gamma_1}$ is a gauge-invariant function on $G^{\Gamma_1}$ and from the bi-invariance of the Haar measure. Indeed, consider a gauge transformation $\phi$ on $\Gamma_1$, which is equal to 1, except maybe on vertices located on $\Gamma_{1,\partial}$. This gauge transformation changes the value of the holonomies along the edges of $\Gamma_{1,\partial}$, and $\phi$ can be chosen in such a way that these holonomies take any prescribed set of values, provided the conjugacy classes of the holonomies along $L_1, \ldots, L_p$ remain unchanged. So, we need just to prove that the value of $I_1$ is preserved by the action of such a gauge transformation. But this action also affects the holonomy along the other edges that meet $\Gamma_{1,\partial}$, so that the gauge invariance of $f_1$ and $D^{\Gamma_1}$ is not sufficient to conclude. It is the bi-invariance of the Haar measure that allows us at this point to forget about the effect of $\phi$ on the edges outside $\Gamma_{1,\partial}$.

Let us write this formally. We set $\Gamma_1 = \{e_1, \ldots, e_k, e_{k+1}, \ldots, e_l, e_{l+1}, \ldots, e_m\}$, where $\Gamma_{1,\partial} = \{e_1, \ldots, e_k\}$ and $e_{k+1}, \ldots, e_l$ correspond to the loops along which the holonomy is imposed by the choice of $y_1$. The point for these edges is that they do not meet $\partial M_1$, so that their discrete holonomy is not affected by $\phi$. For each $i = 1, \ldots, m$, put $g_i^\phi = \phi_{e_i(1)}^{-1} g_i \phi_{e_i(0)}$. We have

$$I_1(g_1, \ldots, g_k) = \int_{G^{\Gamma_1 \setminus \Gamma_{1,\partial}}} f_1 D^{\Gamma_1}(g_1, \ldots, g_m) \, d\nu_{y_1}(g_{k+1}, \ldots, g_l) dg_{l+1} \ldots dg_m$$

$$= \int_{G^{\Gamma_1 \setminus \Gamma_{1,\partial}}} f_1 D^{\Gamma_1}(g_1^\phi, \ldots, g_m^\phi) \, d\nu_{y_1} dg_{l+1} \ldots dg_m$$

$$= \int_{G^{\Gamma_1 \setminus \Gamma_{1,\partial}}} f_1 D^{\Gamma_1}(g_1^\phi, \ldots, g_k^\phi, g_{k+1}, \ldots, g_l, g_{l+1}^\phi, \ldots, g_m^\phi) \, d\nu_{y_1} dg_{l+1} \ldots dg_m$$

$$= \int_{G^{\Gamma_1 \setminus \Gamma_{1,\partial}}} f_1 D^{\Gamma_1}(g_1^\phi, \ldots, g_k^\phi, g_{k+1}, \ldots, g_m) \, d\nu_{y_1} dg_{l+1} \ldots dg_m$$

$$= I_1(g_1^\phi, \ldots, g_k^\phi) = I_1 \circ \phi(g_1, \ldots, g_k).$$

Now let us go back to the computation (5.2). Setting $\Gamma_{2,\partial} = \{e'_1, \ldots, e'_k\}$, in such a way that $e_i$ and $e'_i$ correspond to same edge on $M$, we have

$$\int_{G^{\Gamma_1} \times G^{\Gamma_2}} f_1 f_2 \, dP^{\Gamma_1}_{(y_1, x)} dP^{\Gamma_2}_{(x^{-1}, y')} =$$

$$= \frac{1}{Z_{M_1}(y_1, x) Z_{M_2}(x^{-1}, y_2)} \int_{G^{\Gamma_{1,\partial}} \times G^{\Gamma_{2,\partial}}} I_1(g_1, \ldots, g_k) I_2(g'_1, \ldots, g'_k)$$

$$d\nu_x(g_1, \ldots, g_k) d\nu_{x^{-1}}(g'^{-1}_k, \ldots, g'^{-1}_1)$$

$$= \frac{1}{Z_{M_1}(y_1, x) Z_{M_2}(x^{-1}, y_2)} \int_{G^p} (I_1 I_2)(g_1, \ldots, g_k) \, d\nu_x(g_1, \ldots, g_k)$$

$$= \frac{1}{Z_{M_1}(y_1, x) Z_{M_2}(x^{-1}, y_2)} \int_{G^{\Gamma_\partial}} \left( \int_{G^{\Gamma \setminus \Gamma_\partial}} f_1 D^{\Gamma_1} f_2 D^{\Gamma_2} \, d\nu_{y_1} d\nu_{y_2} dg^{\Gamma'} \right) d\nu_x$$

$$= \frac{Z_M(y_1, x, y_2)}{Z_{M_1}(y_1, x) Z_{M_2}(x^{-1}, y_2)} \int_{G^\Gamma} f_1 f_2 \, dP^\Gamma_{(y_1, x, y_2)}.$$

The equality of the normalization constants follows from the case where $f_1 f_2$ is constant, and this finishes the proof. $\square$

The relation between conditional partition functions that has just been established deserves to be stated separately. We shall discuss this result and another one of the same kind in the second part of this chapter.

PROPOSITION 5.3. *For all $x \in G^p$, $y_1 \in G^{p_1}$ and $y_2 \in G^{p_2}$, the following relation holds:*
$$Z_M(y_1, x, y_2) = Z_{M_1}(y_1, x) Z_{M_2}(x^{-1}, y_2),$$
*where $x$ and $x^{-1}$ refer to the components of $M_1$ and $M_2$ which are identified to build $M$.*

Proposition 5.2 implies very easily Theorem 5.1.

PROOF OF THEOREM 5.1. First, notice that the relation (5.1) implies the conditional independence stated in the first part of the theorem, because $(\mathfrak{x}_1, \ldots, \mathfrak{x}_p) \mapsto P^M_{(\mathfrak{x}_1, \ldots, \mathfrak{x}_p)}$ is a disintegration of the measure $P^M$ with respect to $\mathfrak{H}_{L_1}, \ldots, \mathfrak{H}_{L_p}$. Now, by the continuity of the random holonomy, it is sufficient to prove that this relation holds for functions $f_1$ and $f_2$ that depend on the holonomies along loops that can be put into a graph and Proposition 5.2 applied to the product $f_1 f_2$ tells us exactly that (5.1) holds. □

REMARK 5.4. Consider the simple case of a surface $M$ realized by gluing two surfaces $M_1$ and $M_2$ along a single circle $L$. We take on $L$ the orientation of the boundary of $M_1$. Let $f_1$ be a function on $(\mathcal{M}(PM_1, G), \mathcal{I}_1)$ and $f_2$ a function on $(\mathcal{M}(PM_2, G), \mathcal{I}_2)$. Then $f_1 f_2$ can be seen as a function on $(\mathcal{M}(PM, G), \mathcal{I})$. We have just proved that, for all $\mathfrak{x} \in G/\mathrm{Ad}$, $P^M_{\mathfrak{x}}(f_1 f_2) = P^{M_1}_{\mathfrak{x}}(f_1) P^{M_2}_{\mathfrak{x}^{-1}}(f_2)$. Since $P^M = Z_M^{-1} \int_{G/\mathrm{Ad}} P^M_{\mathfrak{x}} Z_M(\mathfrak{x}) \, d\mathfrak{x}$ (see Theorem 2.72), we get:

$$\begin{aligned} P^M(f_1 f_2) &= \frac{1}{Z_M} \int_{G/\mathrm{Ad}} Z_M(\mathfrak{x}) P^M_{\mathfrak{x}}(f_1 f_2) \, d\mathfrak{x} \\ &= \frac{1}{Z_M} \int_{G/\mathrm{Ad}} Z_{M_1}(\mathfrak{x}) P^{M_1}_{\mathfrak{x}}(f_1) \, Z_{M_2}(\mathfrak{x}^{-1}) P^{M_2}_{\mathfrak{x}^{-1}}(f_2) \, d\mathfrak{x}. \end{aligned}$$

In a more symmetric form, this relation can be written as
$$Z_M P^M(f_1 f_2) = \int_{G/\mathrm{Ad}} Z_{M_1}(\mathfrak{x}) P^{M_1}_{\mathfrak{x}}(f_1) \, Z_{M_2}(\mathfrak{x}^{-1}) P^{M_2}_{\mathfrak{x}^{-1}}(f_2) \, d\mathfrak{x}.$$

The point here is that the analytic objects that glue together in a simple way are not the probability measures, but the measures with their natural weights.

## 5.2. Sewing two surfaces

### 5.2.1. A piece of information missing.
Let us consider again the situation described in the preceding remark, when the two surfaces are glued together along a single circle $L$. In that case, Theorem 5.1 identifies the probability spaces of the Yang-Mills processes on $M_1$ and $M_2$ with two independent subspaces of $(\mathcal{M}(PM, G), \mathcal{I}, P^M_{\mathfrak{x}})$. It is perhaps surprising that these two subspaces do not fill the whole space, in the sense that the inclusion of $\sigma$-fields

$$\mathcal{I}_1 \vee \mathcal{I}_2 \subset \mathcal{I}$$

is *not* an equality in general – in fact when $G$ is not Abelian.

In order to understand this phenomenon, it is convenient to choose a base point $m$ on $L$, say $m = L(0)$ and to consider only loops based at $m$. Recall that, according to Proposition 2.73, the random holonomies along these based loops generate the whole $\sigma$-algebra $\mathcal{I}$.

Let us choose $l_1 \in L_m M_1$ and $l_2 \in L_m M_2$. The random variable $\mathfrak{H}_{l_1,L,l_2}$ is m$\mathcal{I}$ (this stands for "measurable with respect to $\mathcal{I}$"), but is it m$(\mathcal{I}_1 \vee \mathcal{I}_2)$? The random variables $\mathfrak{H}_{l_1,L}$ and $\mathfrak{H}_{L,l_2}$ are respectively m$\mathcal{I}_1$ and m$\mathcal{I}_2$ and we will prove that they provide the whole information about $\mathfrak{H}_{l_1,L,l_2}$ available in $\mathcal{I}_1 \vee \mathcal{I}_2$.

Now the answer to our question depends on the nature of $G$. If it is Abelian, then $\mathfrak{H}_{l_1,L} = (H_{l_1}, H_L)$, $\mathfrak{H}_{L,l_2} = (H_L, H_{l_2})$ and $\mathfrak{H}_{l_1,L,l_2} = (H_{l_1}, H_L, H_{l_2})$ so that $\mathfrak{H}_{l_1,L,l_2}$ is certainly m$(\mathcal{I}_1 \vee \mathcal{I}_2)$. This argument is essentially sufficient to prove that the equality $\mathcal{I}_1 \vee \mathcal{I}_2 = \mathcal{I}$ holds, up to completion with respect to $P_{\mathfrak{x}}^M$.

But if $G$ is not Abelian, the situation is different. Let us take for example $G = SO(3)$. Let us describe concretely the conjugacy and joint conjugacy classes in $SO(3)$ and $SO(3)^n$.

If $r \in SO(3)$ is neither the identity nor a symmetry, it has an angle and an axis, which can be oriented in such a way that the angle is an element of $(0, \pi)$. Let us say that $r$ has a half-axis $u$ which belongs to the unit sphere $S^2$ and an angle $\theta \in (0, \pi)$. This angle characterizes the conjugacy class of $r$ (this is actually still true for the identity and the symmetries). Now, consider $(r_1, \ldots, r_n)$ and $(r'_1, \ldots, r'_n)$ two $n$-tuples of rotations characterized by their half-axes and angles $u_1, \theta_1, u'_1, \theta'_1, \ldots, u_n, \theta_n, u'_n, \theta'_n$. They belong to the same joint conjugacy class if and only if there exists a rotation $R \in SO(3)$ such that $\theta_i = \theta'_i$ and $u'_i = R(u_i)$ for all $i = 1, \ldots, n$.

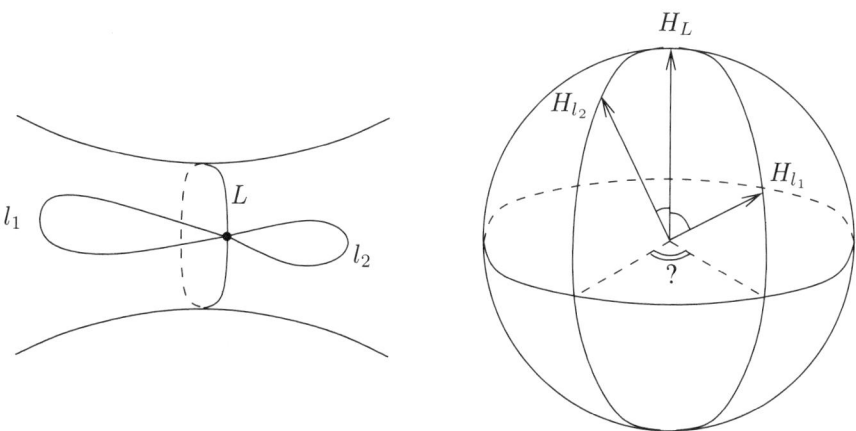

FIGURE 1. Lack of information about the joint class $[H_L, H_{l_1}, H_{l_2}]$ given $[H_L, H_{l_1}]$ and $[H_L, H_{l_2}]$ when $G = SO(3)$.

In our particular case, the random variables $\mathfrak{H}_{l_1,L}$ and $\mathfrak{H}_{L,l_2}$ determine the angles of the three rotations $H_{l_1}, H_L, H_{l_2}$ and the relative position of the half-axes of $H_{l_1}$ and $H_L$ on one hand and that of $H_L$ and $H_{l_2}$ on the other hand. But this is not enough to determine the relative position of the three half-axes (see Figure 1). There remains an undetermined rotation around the axis of $H_L$. Such a rotation is an element of $SO(3)$ which commutes with $H_L$, in other words an element of the centralizer of $H_L$.

This informal argument suggests that $\mathcal{I}_1 \vee \mathcal{I}_2$ is really a smaller $\sigma$-field than $\mathcal{I}$. However, more can be said. If the distributions of $H_{l_1}$ and $H_{l_2}$ are not singular with respect to the Haar measure, which is for example the case if $l_1$ and $l_2$ are non constant simple loops, we claim that the data of $\mathfrak{H}_{l_1,L,l_2}$ allows one to reconstruct the full $\sigma$-field $\mathcal{I}$.

Let us explain this point. Recall that $\mathfrak{x}$ is an element of $SO(3)/\mathrm{Ad}$, which we assume to make things easier is not the identity nor a symmetry, and recall that $\mathfrak{H}_L = \mathfrak{x}$ $P^M_{\mathfrak{x}}$-almost surely. Pick an element $x \in \mathfrak{x}$. Suppose that we are given a measurable section $\tau : (\mathfrak{x} \times SO(3))/\mathrm{Ad} \longrightarrow \{x\} \times SO(3)$ of the canonical projection. A realization of $\mathfrak{H}_{l_1,L,l_2} = [H_{l_1}, H_L, H_{l_2}]$ determines two classes of $(\mathfrak{x} \times SO(3))/\mathrm{Ad}$, namely $[H_L, H_{l_1}]$ and $[H_L, H_{l_2}]$. Let $(x, x_1)$ and $(x, x_2)$ be the images by $\tau$ of these classes. It is easy to check that there exists an element $z$ of the centralizer $C(x)$ of $x$ such that $\mathfrak{H}_{l_1,L,l_2} = [x_1, x, zx_2z^{-1}]$. Observe that the centralizer of $x$ is the subgroup of $SO(3)$ of the rotations with the same axis as $x$. It is isomorphic to $SO(2)$. This element $z$ is determined up to right and left multiplication by an element of $C(x_1) \cap C(x_2)$. Unless the axes of $x_1$ and $x_2$ be equal or opposite to each other, which happens with probability zero if the distributions of $H_{l_1}$ and $H_{l_2}$ have a density, $C(x_1) \cap C(x_2)$ is just $\{1\}$, that is, the center of $SO(3)$ itself. Without using the special features of $SO(3)$, we can say that the coset $zZ(G) \in C(x)/Z(G)$ is well defined.

We have outlined the construction of a random variable $A$ with values in $C(x)/Z(G)$, which is isomorphic in our case to $SO(2)$, and which is nothing but the missing angle of Figure 1. We will prove that it has a uniform distribution, that it is independent of $\mathcal{I}_1 \vee \mathcal{I}_2$ and that $\mathcal{I}_1 \vee \mathcal{I}_2 \vee \sigma(A) = \mathcal{I}$, up to negligible sets. Note that if $G$ is Abelian, the space $C(x)/Z(G) = G/G$ is just a point: there is no contradiction with the fact that $\mathcal{I}_1 \vee \mathcal{I}_2 = \mathcal{I}$.

For the following statement, we choose some elements $\mathfrak{y}_1$ and $\mathfrak{y}_2$ in $(G/\mathrm{Ad})^{p_1}$ and $(G/\mathrm{Ad})^{p_2}$, representing the constraints on the holonomies along other components of the boundaries of $M_1$ and $M_2$ and possibly other loops inside $M_1$ and $M_2$.

THEOREM 5.5. *1. If $G$ is Abelian, then the completions[1] of the $\sigma$-fields $\mathcal{I}_1 \vee \mathcal{I}_2$ and $\mathcal{I}$ with respect to the measure $P^M_{(\mathfrak{y}_1, \mathfrak{x}, \mathfrak{y}_2)}$ are equal.*

*2. If $G$ is not Abelian, there exists a non-trivial sub-$\sigma$-field $\mathcal{A}$ of $\mathcal{I}$ which is independent of $\mathcal{I}_1 \vee \mathcal{I}_2$ and such that the completions of $\mathcal{I}_1 \vee \mathcal{I}_2 \vee \mathcal{A}$ and $\mathcal{I}$ with respect to $P^M_{(\mathfrak{y}_1, \mathfrak{x}, \mathfrak{y}_2)}$ are equal.*

By a non-trivial $\sigma$-field, we mean one that contains non-deterministic events.

There are three steps in the proof of Theorem 5.5: compute the conditional expectation given $\mathcal{I}_1 \vee \mathcal{I}_2$, then construct a random variable $A$ in order to define $\mathcal{A} = \sigma(A)$ and finally show that $\mathcal{A}$ satisfies the claimed properties.

**5.2.2. Conditional expectation.** As a preliminary result, let us show that we can restrict the class of loops that we consider. For this, let us choose on $M$ a Riemannian metric such that $\partial M$ and $L$ are geodesic. Recall that $m$ is the

---

[1] The use of completed $\sigma$-fields is not at all an essential feature of this result: it is imposed by the lack of pathwise regularity of our holonomy process.

basepoint of $L$. The set of piecewise geodesic loops on $M$ based at $m$ is denoted by $\Lambda_m M$.

LEMMA 5.6. *The $\sigma$-field $\mathcal{I}$ is contained in the completion of*
$$\mathcal{J}_{1,2} = \sigma(\mathfrak{H}_{\lambda_1,\ldots,\lambda_n} : \lambda_1, \ldots, \lambda_n \in \Lambda_m M_1 \cup \Lambda_m M_2)$$
*with respect to $P^M_{(\mathfrak{y}_1, \mathfrak{x}, \mathfrak{y}_2)}$.*

PROOF. Recall from Theorem 2.72 that the process $\mathfrak{H}$ satisfies a multiplicativity property which implies in particular that the holonomies along finite products of loops of $\Lambda_m M_1$ and $\Lambda_m M_2$ are $\mathcal{J}_{1,2}$-measurable. According to the continuity property of the random holonomy and to the fact that the holonomy depends only on the equivalence class of the loops, it suffices to show that any loop of $L_m M$ can be approximated by loops that are equivalent to finite products of loops of $\Lambda_m M_1$ and $\Lambda_m M_2$. For this, consider a piecewise geodesic loop of $L_m M$. This loop cuts $L$ transversally at most a finite number of times, hence it is equivalent to a finite product of loops of $\Lambda_m M_1$ and $\Lambda_m M_2$. Since any loop of $L_m M$ can be approximated by piecewise geodesic loops, the result is proved. □

This lemma implies directly the first statement in Theorem 5.5: if $G$ is Abelian, then a random variable of the form $\mathfrak{H}_{\lambda_1,\ldots,\lambda_n}$, with $\lambda_1, \ldots, \lambda_n \in \Lambda_m M_1 \cup \Lambda_m M_2$, is just $(H_{\lambda_1}, \ldots, H_{\lambda_n})$ and this is certainly measurable with respect to $\mathcal{I}_1 \vee \mathcal{I}_2$, because each variable $H_{\lambda_i}$ is either measurable with respect to $\mathcal{I}_1$ or $\mathcal{I}_2$. The result follows immediately.

The case where $G$ is not Abelian is more interesting. Let us compute the conditional distribution given $\mathcal{I}_1 \vee \mathcal{I}_2$ of a random variable $\mathfrak{H}_{\lambda_1,\ldots,\lambda_n}$ as above. For this, it is convenient to introduce the following notation.

Fix $x$ an element of $\mathfrak{x} \in G/\mathrm{Ad}$. Consider an element of $(\mathfrak{x} \times G^n)/\mathrm{Ad}$. It is a class that can be written under the form $[x, x_1, \ldots, x_n]$, and where the $n$-tuple $(x_1, \ldots, x_n)$ is defined up to conjugation by an element of the centralizer $C(x)$. Still, the measure

$$\pi^n_x([x, x_1, \ldots, x_n]) = \int_{C(x)} \delta_{(zx_1 z^{-1}, \ldots, zx_n z^{-1})} \, dz \quad (5.3)$$

is well-defined on $G^n$, and it depends only on the class $[x, x_1, \ldots, x_n]$ and on the choice of $x$. In (5.3), $dz$ denotes the normalized Haar measure on the closed subgroup $C(x)$ of $G$. A short computation shows that for all $y \in G$, $C(\mathrm{Ad}(y)x) = C(yxy^{-1}) = \mathrm{Ad}(y)C(x)$ and for all continuous function $f$ on $G^n$,

$$\int_{G^n} f \, d\pi^n_{\mathrm{Ad}(y)x}([x, x_1, \ldots, x_n]) = \int_{G^n} f \circ \mathrm{Ad}(y) \, d\pi^n_x([x, x_1, \ldots, x_n]). \quad (5.4)$$

PROPOSITION 5.7. *Let $\lambda_1, \ldots, \lambda_n$ be $n$ loops of $\Lambda_m M_1$ and $\lambda'_1, \ldots, \lambda'_{n'}$ $n'$ loops of $\Lambda_m M_2$. For all continuous function $f$ on $G^{n+n'+1}$ invariant by diagonal adjunction we have*

$$E_{P^M_{(\mathfrak{y}_1, \mathfrak{x}, \mathfrak{y}_2)}}[f(\mathfrak{H}_{\lambda_1,\ldots,\lambda_n, L, \lambda'_1,\ldots,\lambda'_{n'}})|\mathcal{I}_1 \vee \mathcal{I}_2] =$$
$$\int_{G^{n+n'+1}} f \, d\pi^n_x(\mathfrak{H}_{L,\lambda_1,\ldots,\lambda_n}) \otimes \delta_x \otimes d\pi^{n'}_x(\mathfrak{H}_{L,\lambda'_1,\ldots,\lambda'_{n'}}) \text{ a.s.}$$

Observe that, by (5.4) and the invariance of $f$, the right hand side depends on $x$ only through its conjugacy class $\mathfrak{x}$.

PROOF. Choose $l_1, \ldots, l_m \in \Lambda_m M_1$ and $l'_1, \ldots, l'_{m'} \in \Lambda_m M_2$. Let $f_1$ be a continuous function on $G^m/\mathrm{Ad} \times G^{m'}/\mathrm{Ad}$. Let $\Gamma$ be a graph on $M$ such that $L$, $l_1, l'_1, \ldots, l_m, l'_m$ and $\lambda_1, \lambda'_1, \ldots, \lambda_n, \lambda'_n$ belong to $\Gamma^*$. Just as in the proof of Theorem 5.1, let us denote by $\Gamma_1$ and $\Gamma_2$ the graphs induced by $\Gamma$ on $M_1$ and $M_2$. Let $f$ be a continuous function on $G^{n+n'+1}$ invariant by diagonal conjugation. By using Proposition 5.2, we find

$$E_{P^M_{(\mathfrak{y}_1, \mathfrak{x}, \mathfrak{y}_2)}}[f(\mathfrak{H}_{\lambda_1,\ldots,\lambda_n,L,\lambda'_1,\ldots,\lambda'_{n'}}) f_1(\mathfrak{H}_{l_1,\ldots,l_m}, \mathfrak{H}_{l'_1,\ldots,l'_{m'}})] =$$

$$= \int_{G^\Gamma} f(h_{\lambda_1},\ldots,h_{\lambda_n},h_L,h_{\lambda'_1},\ldots,h_{\lambda'_{n'}}) f_1([h_{l_1},\ldots,h_{l_m}],[h_{l'_1},\ldots,h_{l'_{m'}}]) \, dP^\Gamma_{(y_1,x,y_2)}$$

$$= \int_{G^{\Gamma_1} \times G^{\Gamma_2}} f(h_{\lambda_1},\ldots,h_{\lambda_n},x,h_{\lambda'_1},\ldots,h_{\lambda'_{n'}}) f_1([h_{l_1},\ldots,h_{l_m}],[h_{l'_1},\ldots,h_{l'_{m'}}])$$
$$dP^{\Gamma_1}_{(y_1,x)} dP^{\Gamma_2}_{(x^{-1},y_2)}.$$

For each element $z \in C(x)$, the gauge transformation equal to 1 at each vertex of $\Gamma_1$ except at $m$ where it takes the value $z$ leaves $P^{\Gamma_1}_{(y_1,x)}$ invariant, and the same property holds for $\Gamma_2$ and $P^{\Gamma_2}_{x^{-1},y_2}$. Thus, the last integral is equal to

$$\int_{G^{\Gamma_1} \times G^{\Gamma_2}} \int_{C(x)^2} f(z_1^{-1} h_{\lambda_1} z_1,\ldots,z_1^{-1} h_{\lambda_n} z_1, x, z_2^{-1} h_{\lambda'_1} z_2,\ldots,z_2^{-1} h_{\lambda'_{n'}} z_2) \, dz_1 dz_2$$
$$f_1([h_{l_1},\ldots,h_{l_m}],[h_{l'_1},\ldots,h_{l'_{m'}}]) \, dP^{\Gamma_1}(y_1,x) dP^{\Gamma_2}(x^{-1},y_2)$$

$$= \int_{G^{\Gamma_1} \times G^{\Gamma_2}} \int_{G^{n+n'+1}} f \, d\pi_x^n([h_L,h_{\lambda_1},\ldots,h_{\lambda_n}]) \otimes \delta_x \otimes d\pi_x^{n'}([h_L,h_{\lambda'_1},\ldots,h_{\lambda'_{n'}}])$$
$$f_1([h_{l_1},\ldots,h_{l_m}],[h_{l'_1},\ldots,h_{l'_{m'}}]) \, dP^{\Gamma_1}(y_1,x) dP^{\Gamma_2}(x^{-1},y_2)$$

$$= E\left[f_1(\mathfrak{H}_{l_1,\ldots,l_m}, \mathfrak{H}_{l'_1,\ldots,l'_{m'}}) \int_{G^{n+n'+1}} f \, d\pi_x^n(\mathfrak{H}_{L,\lambda_1,\ldots,\lambda_n}) \otimes \delta_x \otimes d\pi_x^{n'}(\mathfrak{H}_{L,\lambda'_1,\ldots,\lambda'_{n'}})\right],$$

and this implies the result. $\square$

Proposition 5.7 confirms our intuition that the random variables $\mathfrak{H}_{L,\lambda_1,\ldots,\lambda_n}$ and $\mathfrak{H}_{L,\lambda'_1,\ldots,\lambda'_{n'}}$ contain all the information about $\mathfrak{H}_{\lambda_1,\ldots,\lambda_n,\lambda'_1,\ldots,\lambda'_{n'}}$ available in $\mathcal{I}_1 \vee \mathcal{I}_2$.

We turn now to the construction of a random variable playing the role of the missing angle of Section 5.2.1. All ideas were already explained there, but we need a few technical results.

**5.2.3. Some technical results.** The first point that we need to generalize is the fact that for almost-every pair of rotations $x_1, x_2 \in SO(3)$, one has $C(x_1) \cap C(x_2) = Z(SO(3)) = \{1\}$.

LEMMA 5.8. *Let $G$ be a compact connected Lie group of dimension $n$ and rank $k$. Set $N = n - k + 1$. The set $S_N$ of all $N$-tuples $(g_1, \ldots, g_N)$ such that the closed subgroup of $G$ generated by $g_1, \ldots, g_N$ is $G$ itself has full Haar measure in $G^N$.*

Moreover, $S_N$ is stable by diagonal adjunction, i.e. $(g_1, \ldots, g_N)$ belongs to $S_N$ if and only if $(hg_1 h^{-1}, \ldots, hg_N h^{-1})$ does for all $h$ in $G$.

REMARK 5.9. The value of $N$ given here is not at all the best possible. For example, for $G = SO(3)$, $N = 3$ whereas 2 is enough. What really matters for us is the existence of a finite $N$ for which the property holds.

PROOF. Let us call an element of $G$ *typical* if the closed subgroup that it generates is a maximal torus. The key point is that almost every element of $G$ is typical. To see this, observe first that the set $A$ of typical elements is stable by adjunction. Now fix a maximal torus $T$ in $G$. It is clear that $T$-almost every element of $T$ generates $T$, so that in particular $A \cap T$ has full $T$-measure[2]. The Weyl integration formula, which relates the Haar measures on $T$ and $G$ for central functions, applied to the indicator function of $A$, shows that $A$ has full $G$-measure.

Let $g_1$ be a typical element. It generates a maximal torus $G_1 = T_1$ of $G$ of dimension $k$. If $G_1$ is a proper subgroup of $G$, i.e. if $G$ is not Abelian, then $\dim G_1 < \dim G$ because $G$ is connected. Thus, the complement of $G_1$ has full measure and so does $A \cap G_1^c$. Let $g_2$ be a typical element outside $G_1$, $T_2$ be the maximal torus that it generates and $G_2$ the subgroup generated by $\{g_1, g_2\}$. Let us check that $\dim G_2 > \dim G_1$. If we denote by $\mathfrak{g}_1$, $\mathfrak{g}_2$ and $\mathfrak{t}_2$ the Lie algebras of $G_1$, $G_2$ and $T_2$ respectively, this is a direct consequence of the relations $\mathfrak{t}_2 \not\subset \mathfrak{g}_1$ and $\mathfrak{g}_2 \supset \mathfrak{g}_1 + \mathfrak{t}_2$. By repeating this procedure at most $N = n - k + 1$ times, we get a subgroup $G_N$ which is equal to $G$. It is clear from this construction that the set $S_N$ of convenient $N$-tuples has full Haar measure in $G^N$.

The last part of the statement depends on the fact that two conjugate $N$-tuples generate two conjugate subgroups of $G$. □

In particular, if $(g_1, \ldots, g_N) \in S_N$, then $C(g_1) \cap \ldots \cap C(g_N) = Z(G)$. This is the generalization we were looking for.

The second result is a consequence of a measurable selection theorem proved for example in [17](II, Th. 30, p. 234).

LEMMA 5.10. *There exists a measurable mapping* $\tau : (\mathfrak{x} \times G^N)/\mathrm{Ad} \longrightarrow G^N$ *such that*

$$(\mathfrak{x} \times G^N)/\mathrm{Ad} \longrightarrow \{x\} \times G^N$$
$$[y, y_1, \ldots, y_N] \longmapsto (x, \tau([y, y_1, \ldots, y_N]))$$

*is a measurable section.*

**5.2.4. A piece of information recovered.** Let $N$ be the integer given by Lemma 5.8 and $\tau$ a measurable mapping given by Lemma 5.10. Consider $2N$ simple loops $L_1^-, \ldots, L_N^- \in \Lambda_m M_1$ and $L_1^+, \ldots, L_N^+ \in \Lambda_m M_2$ which meet only at their basepoint $m$. We claim that $\mathfrak{H}_{L_1^-, \ldots, L_N^-}$ and $\mathfrak{H}_{L_1^+, \ldots, L_N^+}$ belong almost surely to $S_N$, which makes sense because $S_N$ is stable under diagonal adjunction. Indeed, one checks easily for example that the distribution of the random variable $(H_{L_1^-}, \ldots, H_{L_N^-})$ on $(\mathcal{M}(PM, G), \mathcal{C})$ has a density with respect to the Haar measure on $G^N$ and the property follows by definition of $N$.

The following result states the existence of a random variable which we claim to be the generalization of our missing angle of Section 5.2.1. Before that, observe

---
[2] By $T$-measure we mean the normalized Haar measure on the Lie group $T$.

that, given any subgroup $H$ of $G$ containing $Z(G)$, the quotient group $H/Z(G)$ exists and acts by adjunction on $G$ and $G^N$.

PROPOSITION 5.11. *There exists on $(\mathcal{M}(PM,G),\mathcal{I})$ a $C(x)/Z(G)$-valued random variable $A$ such that the following equality holds almost surely:*

$$\mathfrak{H}_{L_1^-,\ldots,L_N^-,L,L_1^+,\ldots,L_N^+} = [\tau(\mathfrak{H}_{L,L_1^-,\ldots,L_N^-}), x, \mathrm{Ad}(A)\tau(\mathfrak{H}_{L,L_1^+,\ldots,L_N^+})]. \qquad (5.5)$$

PROOF. Fix $\omega$ in the event where $\mathfrak{H}_{L_1^-,\ldots,L_N^-}$ and $\mathfrak{H}_{L_1^+,\ldots,L_N^+}$ belong to $S_N$. Define $x^- = (x_1^-,\ldots,x_N^-)$ and $x^+ = (x_1^+,\ldots,x_N^+)$ in $G^N$ by

$$x^- = \tau(\mathfrak{H}_{L,L_1^-,\ldots,L_N^-}(\omega)), \quad x^+ = \tau(\mathfrak{H}_{L,L_1^+,\ldots,L_N^+}(\omega)).$$

Since $[x^-, x] = \mathfrak{H}_{L_1^-,\ldots,L_N^-,L}$ and $[x, x^+] = \mathfrak{H}_{L,L_1^+,\ldots,L_N^+}$, one checks easily that there exists an element $z$ of $C(x)$ such that

$$[x^-, x, zx^+z^{-1}] = \mathfrak{H}_{L_1^-,\ldots,L_N^-,L,L_1^+,\ldots,L_N^+}. \qquad (5.6)$$

If another element $z'$ of $C(x)$ satisfies the same relation, it is readily checked that there exist $w_1 \in C(x_1^-) \cap \ldots \cap C(x_N^-) = Z(G)$ and $w_2 \in C(x_1^+) \cap \ldots \cap C(x_N^+) = Z(G)$ such that $z' = w_1 z w_2 = z w_1 w_2$. Conversely, any $z'$ of the form $zw$ with $w \in Z(G)$ could replace $z$ in (5.6).

Thus the class $zZ(G)$ is well-defined in $C(x)/Z(G)$ and we set

$$A(\omega) = zZ(G).$$

That (5.5) holds with this definition follows from (5.6). □

REMARK 5.12. In the proof above, $x^-$ and $x^+$ are measurable functions of $\mathfrak{H}_{L_1^-,\ldots,L_N^-,L,L_1^+,\ldots,L_N^+}$ so that $A$ itself can be written as $\tilde{A}(\mathfrak{H}_{L_1^-,\ldots,L_N^-,L,L_1^+,\ldots,L_N^+})$ for some measurable function $\tilde{A}$. Properly speaking, this function is defined almost everywhere on $(G^N \times \mathfrak{x} \times G^N)/\mathrm{Ad}$ and one checks that, for all $z_-, z_+$ in $C(x)$ and all $x^-, x^+$ in $G^N$,

$$\tilde{A}([z_-x^-z_-^{-1}, x, z_+x^+z_+^{-1}]) = z_-^{-1}z_+ \tilde{A}([x^-, x, x^+]). \qquad (5.7)$$

In the following result, we speak about uniform distribution on $C(x)/Z(G)$. This makes sense because, as $Z(G)$ is a closed normal subgroup of $C(x)$, the quotient $C(x)/Z(G)$ is still a compact Lie group on which we may consider the normalized Haar measure.

PROPOSITION 5.13. *The random variable $A$ is uniformly distributed on the Lie group $C(x)/Z(G)$ and independent of $\mathcal{I}_1 \vee \mathcal{I}_2$ under the measure $P^M_{(\mathfrak{y}_1,\mathfrak{x},\mathfrak{y}_2)}$.*

PROOF. Let $f$ be a continuous function on $C(x)/Z(G)$. By using Proposition 5.7, we find

$$E[f(A)|\mathcal{I}_1 \vee \mathcal{I}_2] = E_{P^M_{(\mathfrak{y}_1,\mathfrak{x},\mathfrak{y}_2)}}[f(\tilde{A}(\mathfrak{H}_{L_1^-,\ldots,L_N^-,L,L_1^+,\ldots,L_N^+}))|\mathcal{I}_1 \vee \mathcal{I}_2]$$

$$= \int_{G^{2N+1}} f \circ \tilde{A} \, d\pi_x^N(\mathfrak{H}_{L,L_1^-,\ldots,L_N^-}) \otimes \delta_x \otimes d\pi_x^N(\mathfrak{H}_{L,L_1^+,\ldots,L_N^+}).$$

Set $x^- = \tau(\mathfrak{H}_{L,L_1^-,\ldots,L_N^-})$, $x^+ = \tau(\mathfrak{H}_{L,L_1^+,\ldots,L_N^+})$. By definition of $\pi_x^N$ and thanks to (5.7), we have

$$E[f(A)|\mathcal{I}_1 \vee \mathcal{I}_2] = \int_{C(x)^2} f(\tilde{A}([z_-x^-z_-^{-1}, x, z_+x^+z_+^{-1}]))\, dz_-dz_+$$

$$= \int_{C(x)^2} f(z_-^{-1}z_+\tilde{A}([x^-, x, x^+]))\, dz_-dz_+$$

$$= \int_{C(x)} f(z)\, dz.$$

Since this relation holds for all continuous $f$, the result is proved. □

PROOF OF THEOREM 5.5. If $G$ is Abelian, we have already noticed that the result follows from Lemma 5.6.

If $G$ is not Abelian, let us define $\mathcal{A} = \sigma(A)$. By Proposition 5.13, it is independent of $\mathcal{I}_1 \vee \mathcal{I}_2$, and it is non-trivial, because $Z(G)$ is always a proper subgroup of $C(x)$. To see this last point, one can split the Lie algebra of $G$ as $\mathfrak{g} = \mathfrak{z} \oplus [\mathfrak{g}, \mathfrak{g}]$ where $\mathfrak{z}$ is the center of $\mathfrak{g}$. If $x = \exp X$, let $X = X_\mathfrak{z} + X'$ be the corresponding decomposition. If $X' = 0$, then $x$ belongs to the center of $G$ and $C(x) = G$ contains strictly $Z(G)$. Otherwise, $C(x)$ contains $\exp(\mathfrak{z} \oplus \mathbb{R}X')$ and its dimension is larger than that of $\mathfrak{z}$, which is also that of $Z(G)$.

There remains to show that $\mathcal{I}_1 \vee \mathcal{I}_2 \vee \mathcal{A}$ contains $\mathcal{J}_{1,2}$. The relation (5.5) implies that the random variable $\mathfrak{H}_{L_1^-,\ldots,L_N^-,L,L_1^+,\ldots,L_N^+}$ is measurable with respect to $\mathcal{I}_1 \vee \mathcal{I}_2 \vee \mathcal{A}$. It is sufficient now to prove that this variable, together with $\mathcal{I}_1$ and $\mathcal{I}_2$, generates $\mathcal{J}_{1,2}$. For this, pick $p$ loops $\lambda_1, \ldots, \lambda_p$ in $\Lambda_m M_1$ and $q$ loops $\lambda_1', \ldots, \lambda_q'$ in $\Lambda_m M_2$. We claim that the values of the three variables $\mathfrak{H}_{\lambda_1,\ldots,\lambda_p,L_1^-,\ldots,L_N^-}$, $\mathfrak{H}_{L_1^-,\ldots,L_N^-,L_1^+,\ldots,L_N^+}$ and $\mathfrak{H}_{L_1^+,\ldots,L_N^+,\lambda_1',\ldots,\lambda_q'}$ determine that of $\mathfrak{H}_{\lambda_1,\ldots,\lambda_p,\lambda_1',\ldots,\lambda_q'}$.

To see this, take $x^-$ in $G^N$ such that $[x^-] = \mathfrak{H}_{L_1^-,\ldots,L_N^-}$. Almost surely, there is a unique choice of $u$ in $G^p$, $x^+$ in $G^N$, then $u'$ in $G^q$, such that $[u, x^-]$, $[x^-, x^+]$ and $[x^+, u']$ are equal to the values of the three random variables mentioned above, in the same order. The class $[u, u']$ does not depend on the choice of $x^-$ and it is exactly the value of $\mathfrak{H}_{\lambda_1,\ldots,\lambda_p,\lambda_1',\ldots,\lambda_q'}$, as one can check directly from the definition of this variable on the space $\mathcal{M}(PM, G)$. □

### 5.3. Mending a surface

It is possible to prove results similar to Theorems 5.1 and 5.5 in the situation where a surface $M$ is obtained by sewing together two components of the boundary of a single surface $M_1$. There is of course no more conditional independence in this situation, but it is still possible to study the relationship between the Yang-Mills measures on $M$ and $M_1$. This is the purpose of this section.

#### 5.3.1. Comparison of the probability spaces.
Let $M_1$ be an oriented surface with boundary such that $\partial M_1$ has at least two components $N_1$ and $N_2$. Let $\psi : N_1 \longrightarrow N_2$ be an orientation-reversing diffeomorphism and let $M$ be obtained by gluing $M_1$ with itself along $\psi$. Let $L_1$ and $L_2$ be two loops on $M_1$ whose images are $N_1$ and $-N_2$ respectively and call $L$ the corresponding loop on $M$. Set $m = L(0)$, $m_1 = L_1(0)$ and $m_2 = L_2(0)$. Note that, in contrast to the preceding situation, $M_1$

is not embedded in $M$, it is only immersed and we denote by $i : M_1 \longrightarrow M$ this immersion. It extends to a mapping from $LM_1$ to $LM$ whose restriction to $L_{m_1}M_1$ is injective. Since the $\sigma$-field $\mathcal{I}_1$ on $\mathcal{M}(PM_1, G)$ is generated by the variables $\mathfrak{H}_{l_1,\ldots,l_n}$ with $l_1, \ldots, l_n \in L_{m_1}M_1$ (see Proposition 2.73), we identify it with the sub-$\sigma$-field $\sigma(\mathfrak{H}_{l_1,\ldots,l_n} : l_1, \ldots, l_n \in i(L_{m_1}M_1))$ of $\mathcal{I}$.

THEOREM 5.14. *Let $f$ be a measurable function on $(\mathcal{M}(PM_1, G), \mathcal{I}_1)$. Then $f$ can be seen as a random variable on $(\mathcal{M}(PM, G), \mathcal{I})$ and, for all $\mathfrak{x}$ in $G/\mathrm{Ad}$, the following equality holds:*

$$P_{\mathfrak{x}}^M(f) = P_{(\mathfrak{x}, \mathfrak{x}^{-1})}^{M_1}(f), \qquad (5.8)$$

*where the condition is put on $\mathfrak{H}_L$ in the left-hand side and on the holonomies along $N_1$ and $N_2$ in the right-hand side.*

*Moreover, this remains true with the obvious modifications if we condition further the Yang-Mills measures with respect to the random holonomy along some other boundary components or some interior loops on $M_1$.*

As for the proof of Theorem 5.1, we begin with a discrete result. Given a graph $\Gamma$ on $M$, there exists a graph $\Gamma_1$ on $M_1$ such that $\Gamma_1$ is mapped by $i$ to $\Gamma$ and such that the edges of $\Gamma_1$ lying on $N_1$ and $N_2$ respectively are in one-to-one correspondence via the diffeomorphism $\psi$. We call $\Gamma_1$ the graph *induced* by $\Gamma$ on $M_1$.

PROPOSITION 5.15. *Let $\Gamma$ be a graph on $M$ and $\Gamma_1$ the graph induced on $M_1$ by $\Gamma$. Let $f$ be a continuous function on $G^{\Gamma_1}$ which is invariant by the gauge transformations $\phi^1$ such that $\phi_{m_1}^1 = \phi_{m_2}^1 = 1$. Then $f$ gives rise to a function on $G^{\Gamma}$ invariant by the gauge transformations $\phi$ such that $\phi_m = 1$ and, for all $x \in G$,*

$$\int_{G^{\Gamma_1}} f \, dP_{(y,x,x^{-1})}^{\Gamma_1} = \int_{G^{\Gamma}} f \, dP_{(y,x)}^{\Gamma}, \qquad (5.9)$$

*where $x$ is the holonomy along $N_1$ and $-N_2$ and $y$ stands for the values of the holonomies along the other components of $\partial M_1$ and possibly other loops inside $M_1$.*

The proof of this proposition is similar to the proof of Proposition 5.2, with a small difference because we consider special gauge transformations. Note that Proposition 5.15 implies the corresponding statement without any restriction on the gauge transformations, which is in fact enough to prove Theorem 5.14. However, we will need this refinement in the proof of Proposition 5.21.

PROOF OF PROPOSITION 5.15. We have

$$\int_{G^{\Gamma_1}} f \, dP_{(y,x,x^{-1})}^{\Gamma_1} = \frac{1}{Z_{M_1}(y,x,x^{-1})} \int_{G^{\Gamma_1}} fD^{\Gamma_1} \, d\nu_y d\nu_x d\nu_{x^{-1}} dg'$$

$$= \frac{1}{Z_{M_1}(y,x,x^{-1})} \int_{G^{\Gamma_{1,\partial}}} d\nu_x d\nu_{x^{-1}} \int_{G^{\Gamma_1 \setminus \Gamma_{1,\partial}}} fD^{\Gamma_1} \, d\nu_y dg',$$

where $\Gamma_{1,\partial}$ denotes the restriction of $\Gamma_1$ to $\partial M_1$. The last integral is a function on $G^{\Gamma_{1,\partial}}$ and the same argument as in the proof of Proposition 5.2 together with the fact that we consider only gauge transformations such that $\phi_{m_1}^1 = \phi_{m_2}^1 = 1$ shows that this function depends only on the values of $h_{L_1}$ and $h_{L_2}$. Thus, we can drop the integration against $d\nu_{x^{-1}}$ and replace the variables associated with the edges lying on $N_2$ by the variables associated to the corresponding edges on $N_1$. This gives

$$\int_{G^{\Gamma_1}} f \, dP^{\Gamma_1}_{(\mathfrak{y},\mathfrak{x},\mathfrak{x}^{-1})} = \frac{1}{Z_{M_1}(\mathfrak{y},\mathfrak{x},\mathfrak{x}^{-1})} \int_{G^{\Gamma_\partial}} d\nu_x \int_{G^{\Gamma \setminus \Gamma_\partial}} f D^\Gamma \, d\nu_y dg'$$

$$= \frac{Z_M(\mathfrak{y},\mathfrak{x})}{Z_{M_1}(\mathfrak{y},\mathfrak{x},\mathfrak{x}^{-1})} \int_{G^\Gamma} f \, dP^\Gamma_{(\mathfrak{y},\mathfrak{x})}.$$

The equality of the normalization constants follows from the case where $f$ is constant and this finishes the proof. □

Once again, we state separately the property of the conditional partition functions that we have just established.

PROPOSITION 5.16. *For all $\mathfrak{x}_1, \ldots, \mathfrak{x}_p, \mathfrak{x} \in G/\mathrm{Ad}$, the following equality holds:*

$$Z_M(\mathfrak{x}_1, \ldots, \mathfrak{x}_p, \mathfrak{x}, \mathfrak{x}^{-1}) = Z_{M_1}(\mathfrak{x}_1, \ldots, \mathfrak{x}_p, \mathfrak{x}),$$

*where $\mathfrak{x}$ and $\mathfrak{x}^{-1}$ are the holonomies along the components of $\partial M_1$ which we glue together to obtain $M$.*

PROOF OF THEOREM 5.14. The proof is similar to that of Theorem 5.1: by the continuity of the random holonomy, it is enough to consider the case of a cylindrical function $f$ which depends on the holonomy along loops that can be put into a graph. For such a function, (5.8) is equivalent to (5.9). □

**5.3.2. Defect of information.** Let us begin by a short informal discussion. There are two pieces of information which are present in $\mathcal{I}$ but apparently not in $\mathcal{I}_1$. The first one corresponds to the holonomy along loops based at $m$ but lying on the wrong side of $L$, that is, which cannot be lifted in $L_{m_1} M_1$. Even worse, the second one comes from the fact that any open path from $m_1$ to $m_2$ in $M_1$ is sent to a loop on $M$ by the immersion $i$. As we shall see, it is sufficient to recover the second bit of information to fill the gap between $\mathcal{I}_1$ and $\mathcal{I}$.

For the moment, let us choose a path $R$ from $m_1$ to $m_2$ on $M_1$. The holonomy along $i(R)$ is a random variable on $\mathcal{M}(PM, G)$ but certainly not a $\mathcal{I}_1$-measurable one. Once again, what is missing is closely related to the centralizer of $H_L$. To see why, let us consider the random variable $\mathfrak{H}_{L_1, RL_2R^{-1}}$. Since the holonomies along $L_1$ and $L_2$ must be equal after the gluing procedure, let us choose arbitrarily an element $x \in \mathfrak{x}$ and compute as if $H_{L_1} = H_{L_2} = x$, although this does not make sense in $\mathcal{I}_1$. The variable $\mathfrak{H}_{L_1, RL_2R^{-1}}$, which is $\mathcal{I}_1$-measurable, determines to some extent the holonomy along $R$: indeed, the class $[x, H_R^{-1} x H_R]$ determines the value of $H_R$ up to left and right multiplication by elements of the centralizer $C(x)$.

The main result is similar to Theorem 5.5.

THEOREM 5.17. *There exists a non-trivial sub-$\sigma$-algebra $\mathcal{A}'$ in $\mathcal{I}$ which is independent of $\mathcal{I}_1$ and such that the completions of $\mathcal{I}_1 \vee \mathcal{A}'$ and $\mathcal{I}$ with respect to $P^M_{(\mathfrak{y},\mathfrak{x})}$ are equal.*

REMARK 5.18. The conclusion of this theorem does not depend on the fact that $G$ is Abelian or not. In particular, $\mathcal{A}'$ is non-trivial even when $G$ is Abelian.

The proof of this theorem has very much the same structure as the proof of Theorem 5.5. We begin with a lemma which allows us to consider a restricted class of loops. For this, let us endow $M$ with a Riemannian metric such that $L$ is geodesic. This metric on $M$ induces by pull-back by $i$ a metric on $M_1$. We fix once for all a piecewise geodesic path $R$ in $M$ joining $m_1$ to $m_2$ and meeting $\partial M_1$ only at $m_1$ and $m_2$. Recall that $\Lambda_{m_1} M_1$ denotes the set of piecewise geodesic loops on $M_1$ based at $m_1$.

LEMMA 5.19. *The $\sigma$-field $\mathcal{I}$ is contained in the completion of*
$$\mathcal{J}_1 = \sigma(\mathfrak{H}_{\lambda_1,\ldots,\lambda_n}, \lambda_1, \ldots, \lambda_n \in i(\Lambda_{m_1} M_1 \cup \{R\}))$$
*with respect to the measure* $P^M_{(\mathfrak{y},\mathfrak{x})}$.

REMARK 5.20. This lemma gives a rigorous content to the claim made at the beginning of this section, namely that the information corresponding to the loops lying on the wrong side of $L$ in $M$ can be recovered once one knows the holonomy along a path from $m_1$ to $m_2$.

PROOF. Just as in the proof of Lemma 5.6, the point is to prove that any loop of $L_m M$ can be approximated by loops that are equivalent to finite products of loops of $i(\Lambda_{m_1} M_1 \cup \{R\})$. The result follows then by the multiplicativity of $\mathfrak{H}$.

Consider a piecewise geodesic loop. Since it cuts at most a finite number of times $L$ transversally, it is equivalent to a finite product of loops of $\Lambda_m M$ that are the images by $i$ of loops based at $m_1$ or at $m_2$ or of paths with endpoints $m_1$ and $m_2$. Conjugation by $R$ or left or right multiplication by $R$ transform all these paths on $M_1$ into loops based at $m_1$, and expressing the holonomy along our piecewise geodesic loop amounts to expressing that along the loops of $i(\Lambda_{m_1} M_1 \cup \{R\})$. As usual, we conclude by using the fact that any loop of $L_m M$ is a limit in length of piecewise geodesic loops. □

Now, let us compute the conditional expectation of a variable $\mathfrak{H}_{\lambda_1,\ldots,\lambda_n}$ with respect to $\mathcal{I}_1$. For this, we introduce a probability measure similar to $\pi_x^n$ defined in the case of two surfaces.

Observe that the mapping $\Lambda_{m_1} M_1 \cup \{R\} \longrightarrow \Lambda_m M$ induced by $i$ is injective, so that we can regard $\Lambda_{m_1} M_1 \cup \{R\}$ as a subset of $\Lambda_m M$.

Let us fix an element $x$ in $\mathfrak{x}$ and consider an element of $(\mathfrak{x} \times \mathfrak{x} \times G^n)/\mathrm{Ad}$. It is a joint conjugacy class which can be written $[x, x_R x x_R^{-1}, x_1, \ldots, x_n]$ and we leave it to the reader to check that for any other representation $[x, y_R x y_R^{-1}, y_1, \ldots, y_n]$ of the same class, there exist two elements $z_1$ and $z_2$ in the centralizer of $x$ such that $(y_1, \ldots, y_n) = \mathrm{Ad}(z_1)(x_1, \ldots, x_n)$ and $y_R = z_1 x_R z_2$. Thus, the measure

$$\alpha_x^{n+1}([x, x_R x x_R^{-1}, x_1, \ldots, x_n]) = \int_{C(x)^2} \delta_{z_1 x_1 z_1^{-1}, \ldots, z_1 x_n z_1^{-1}, z_1 x_R z_2} \, dz_1 dz_2$$

is well-defined on $G^{n+1}$ and depends only on $x$ and the class $[x, x_R^{-1} x x_R, x_1, \ldots, x_n]$. Observe that, for all function $f$ invariant by diagonal adjunction, one has

$$\int_{G^{n+1}} f \, d\alpha_x^{n+1}([x, x_R x x_R^{-1}, x_1, \ldots, x_n]) = \int_{C(x)} f(x_1, \ldots, x_n, x_R z) \, dz.$$

PROPOSITION 5.21. *Let $\lambda_1, \ldots, \lambda_n$ be $n$ loops of $\Lambda_{m_1} M_1$. For all continuous function $f$ on $G^{n+1}$ invariant by diagonal adjunction, we have*

$$E[f(\mathfrak{H}_{\lambda_1,\ldots,\lambda_n,R})|\mathcal{I}_1] = \int_{G^{n+1}} f \, d\alpha_x^{n+1}(\mathfrak{H}_{L, RLR^{-1}, \lambda_1, \ldots, \lambda_n}).$$

PROOF. Choose $l_1, \ldots, l_m$ in $\Lambda_{m_1} M_1$ and a continuous function $f_1$ on $G^m/\mathrm{Ad}$. Let $\Gamma$ be a graph on $M$ such that all the loops we are considering belong to $\Gamma^*$. Denote by $\Gamma_1$ the graph induced by $\Gamma$ on $M_1$, in the same way as we did in the proof of Theorem 5.14.

$$E_{P_{\mathfrak{y},\mathfrak{x}}^M}[f(\mathfrak{H}_{\lambda_1,\ldots,\lambda_n,R})f_1(\mathfrak{H}_{l_1,\ldots,l_m})] = \int_{G^\Gamma} f(h_{\lambda_1},\ldots,h_{\lambda_n},h_R)f_1(h_{l_1},\ldots,h_{l_m})\,dP_{(y,x)}^\Gamma$$

$$= \int_{G^{\Gamma_1}} f(h_{\lambda_1},\ldots,h_{\lambda_n},h_R)f_1(h_{l_1},\ldots,h_{l_m})\,dP_{(y,x,x^{-1})}^{\Gamma_1},$$

thanks to Proposition 5.15. It should be noted that the integrand is not gauge-invariant because of the presence of $h_R$, but it is precisely invariant by those gauge transformations $\phi$ such that $\phi_{m_1} = \phi_{m_2} = 1$.

We are now working on $M_1$, where the endpoints of $R$ are different and we can use the invariance of $P_{(y,x,x^{-1})}^{\Gamma_1}$ under the gauge transformation identically equal to 1, except at $m_1$ where it takes the value $z \in C(x)$. This transformation leaves the $h_{\lambda_i}$'s and $h_{l_i}$'s invariant since they are based at $m_1$, and transforms $h_R$ into $h_R z$. By doing this for all $z \in C(x)$ and integrating with respect to the uniform measure on $C(x)$, we find

$$E_{P_{\mathfrak{y},\mathfrak{x}}^M}[f(\mathfrak{H}_{\lambda_1,\ldots,\lambda_n,R})f_1(\mathfrak{H}_{l_1,\ldots,l_m})] =$$

$$= \int_{G^{\Gamma_1}} \left[\int_{C(x)} f(h_{\lambda_1},\ldots,h_{\lambda_n},h_R z)\,dz\right] f_1(h_{l_1},\ldots,h_{l_m})\,dP_{(y,x,x^{-1})}^{\Gamma_1}$$

$$= \int_{G^{\Gamma_1}} \left[\int_{G^{n+1}} f\,d\alpha_x^{n+1}([h_L, h_{RLR^{-1}}, h_{\lambda_1},\ldots,h_{\lambda_n}])\right] f_1(h_{l_1},\ldots,h_{l_m})\,dP_{(y,x,x^{-1})}^{\Gamma_1}$$

$$= E\left[\left(\int_{G^{n+1}} f\,d\alpha_x^{n+1}(\mathfrak{H}_{L,RLR^{-1},\lambda_1,\ldots,\lambda_n})\right) f_1(\mathfrak{H}_{l_1,\ldots,l_m})\right],$$

and this is exactly the result. □

We follow the proof of Theorem 5.5 and construct now a random variable $A'$ that will generate the $\sigma$-algebra $\mathcal{A}'$. We will use the existence of the integer $N$ and a measurable section $\tau$ defined by Lemmas 5.8 and 5.10. We need also a new mapping. Recall that $x$ is still fixed in $\mathfrak{x}$.

LEMMA 5.22. *There exists a measurable mapping $\sigma : \mathfrak{x} \longrightarrow G$ such that, for all $y \in \mathfrak{x}$,*

$$y = \mathrm{Ad}(\sigma(y))x.$$

The existence of such a measurable section is established by using the result of [17] mentioned before Lemma 5.10. We choose $\sigma$ once for all.

Let us fix now $N$ simple loops $L_1, \ldots, L_N$ in $\Lambda_{m_1} M_1$ meeting only at their basepoint, so that $\mathfrak{H}_{L_1,\ldots,L_N}$ belongs $P_{(\mathfrak{y},\mathfrak{x})}^M$-almost surely to $S_N$.

PROPOSITION 5.23. *There exists on $(\mathcal{M}(PM, G), \mathcal{I})$ a $C(x)$-valued random variable such that the following equality holds $P_{(\mathfrak{y},\mathfrak{x})}^M$-almost surely:*

$$\mathfrak{H}_{L,L_1,\ldots,L_N,R} = [\tau(\mathfrak{H}_{L,L_1,\ldots,L_N}), \sigma(\mathfrak{H}_{RLR^{-1}})A']. \tag{5.10}$$

PROOF. Let $\omega$ be such that $\mathfrak{H}_{L_1,\ldots,L_N}$ belongs to $S_N$. Set $(x_1,\ldots,x_N) = \tau(\mathfrak{H}_{L,L_1,\ldots,L_N})$ and choose $x_R \in G$ such that

$$\mathfrak{H}_{L,L_1,\ldots,L_N,R}(\omega) = [x, x_1, \ldots, x_N, x_R]. \tag{5.11}$$

Then $x_R$ is defined up to conjugation by an element of $C(x_1) \cap \ldots \cap C(x_N) = Z(G)$, so that it is in fact uniquely defined.

Now, set $x'_R = \sigma(x_R x x_R^{-1})$. The product $x'_R{}^{-1} x_R$ belongs to $C(x)$ and we set
$$A'(\omega) = x'_R{}^{-1} x_R.$$
With this definition, (5.10) follows immediately from (5.11). □

REMARK 5.24. The random variable $A'$ depends on $\omega$ through $\mathfrak{H}_{L,L_1,\ldots,L_N,R}(\omega)$. Thus, there exists a measurable function $\tilde{A}'$ such that $A' = \tilde{A}'(\mathfrak{H}_{L,L_1,\ldots,L_N,R})$. This function $\tilde{A}'$ is defined almost everywhere on $(\mathfrak{x} \times G^N \times \mathfrak{x})/\mathrm{Ad}$ and it is easily checked that, for all $z \in C(x)$,
$$\tilde{A}'([x, x_1, \ldots, x_N, x_R z]) = \tilde{A}'([x, x_1, \ldots, x_N, x_R])z. \quad (5.12)$$

PROPOSITION 5.25. *The random variable $A'$ is independent of $\mathcal{I}_1$ and uniformly distributed on $C(x)$ under the measure $P^M_{(\mathfrak{y},\mathfrak{x})}$.*

PROOF. Let $f$ be a continuous function on $C(x)$. By using Proposition 5.21, we get
$$\begin{aligned} E[f(A)|\mathcal{I}_1] &= E[f(\tilde{A}'(\mathfrak{H}_{L,L_1,\ldots,L_N,R}))|\mathcal{I}_1] \\ &= \int_{G^{N+2}} f \circ \tilde{A}' \, d\alpha_x^{N+2}(\mathfrak{H}_{L,RLR^{-1},L,L_1,\ldots,L_N}). \end{aligned}$$

Let us fix $\omega$ and set $(x, x_1, \ldots, x_N) = \tau(\mathfrak{H}_{L,L_1,\ldots,L_N}(\omega))$. We assume that $C(x_1) \cap \ldots \cap C(x_N) = Z(G)$. Let $x_R$ be the unique element of $G$ such that $\mathfrak{H}_{L,L_1,\ldots,L_N,R}(\omega) = [x, x_1, \ldots, x_N, x_R]$. Then, in particular, $\mathfrak{H}_{L,RLR^{-1},L,L_1,\ldots,L_N} = [x, x_R^{-1} x x_R, x_1, \ldots, x_N]$, so that, by definition of $\alpha_x^{N+2}$ and thanks to (5.12),
$$\begin{aligned} E[f(A')|\mathcal{I}_1](\omega) &= \int_{C(x)} f(\tilde{A}'(x, x_1, \ldots, x_N, x_R z)) \, dz \\ &= \int_{C(x)} f(\tilde{A}'(x, x_1, \ldots, x_N, x_R)z) \, dz \\ &= \int_{C(x)} f(z) \, dz. \end{aligned}$$

The result follows immediately. □

PROOF OF THEOREM 5.17. Set $\mathcal{A}' = \sigma(A')$. By Proposition 5.25, we know that $\mathcal{A}'$ is independent of $\mathcal{I}_1$ and also that it is non trivial, since the centralizer of any element contains at least a one-parameter subgroup of $G$.

By (5.10), $\mathfrak{H}_{L,L_1^1,\ldots,L_N^1,c}$ is measurable with respect to $\mathcal{I}_1$ so that we need only prove that $\mathcal{I}_1 \vee \sigma(\mathfrak{H}_{L,L_1^1,\ldots,L_N^1,c})$ contains $\mathcal{J}_1$ to finish the proof.

Let $\lambda_1, \ldots, \lambda_p$ be $p$ loops of $i(\Lambda_{m_1} M_1 \cup \{R\})$. We need to prove that the values of the variables $\mathfrak{H}_{R,L_1,\ldots,L_N}$ and $\mathfrak{H}_{L_1,\ldots,L_N,\lambda_1,\ldots,\lambda_p}$ determine that of $\mathfrak{H}_{R,\lambda_1,\ldots,\lambda_p}$ almost surely. This is done along the same lines as in the proof of Theorem 5.5. □

REMARK 5.26. When $G$ is Abelian, $C(x)$ is $G$ itself, so that the theorem says that it is necessary to add an independent uniform random variable to the holonomy process in order to close a handle. This is not surprising in the light of the study made in Chapter 3, where it has appeared that, when $G = U(1)$, the holonomies along a set of generators of the homology are uniform, independent and independent of the holonomies along homologically trivial loops.

## 5.4. An area-dependent topological field theory

We have not emphasized the role of the conditional partition functions in the development of the theory but the reader will have noticed that the proof of each fundamental result has led, as a byproduct, to a non-trivial property of these functions. Let us remind the reader of these properties. The gauge-invariance of the discrete Yang-Mills measure (Proposition 1.18) has given the invariance by conjugation of the partition function (Proposition 1.19). Then, the invariance under subdivision of the discrete measure (Theorem 1.22) has given the fact that the partition function is invariant under refinement of the graph (Proposition 1.26). The identification of the discrete and continuous random holonomies (Proposition 2.46) has led to the complete invariance with respect to the graph (Proposition 2.50). Finally, the discrete Markov property (Proposition 5.2) and the description of the modification of the discrete measure when one glues two components of the boundary together (Proposition 5.14) have led respectively to Propositions 5.3 and 5.16, which we are going to study more closely now.

This indicates that the partition functions are closely related to several aspects of the theory. We are going to summarize the properties of the conditional partition functions and this will reveal the nice algebraic structure of this set of functions. The computations presented here are in the same vein as those of [**44**].

### 5.4.1. Algebraic properties of the partition functions.
Let $(M, \sigma)$ be a surface, with a boundary $\partial M = N_1 \cup \ldots \cup N_p$ or without boundary. For each graph $\Gamma$ on $M$ and any $x_1, \ldots, x_p \in G$, the number $\int_{G^\Gamma} D^\Gamma \, d\nu_{x_1} \ldots d\nu_{x_p} \, dg'$ is well defined (see Section 1.5). By Lemma 1.19, it depends only on the conjugacy classes of the $x_i$'s: it is a central function of the $x_i$'s. By Proposition 2.50, which is true on a surface with boundary by Lemma 2.55, this number is independent of $\Gamma$. If $\mathfrak{x}_1, \ldots, \mathfrak{x}_p$ are the conjugacy classes of $x_1, \ldots, x_p$, we denote this number by $Z_M(\mathfrak{x}_1, \ldots, \mathfrak{x}_p)$, or just $Z_M$ if $M$ is closed.

Consider now an area-preserving diffeomorphism between $(M, \sigma)$ and another surface $(M', \sigma')$, that is, a diffeomorphism between $M$ and $M'$ which sends $\sigma$ to $\sigma'$. Then $\int_{G^\Gamma} D^\Gamma \, d\nu_{x_1} \ldots d\nu_{x_p} \, dg'$ is obviously invariant by this diffeomorphism. Thus, the function $Z_M$ depends on $M$ only through its class modulo area-preserving diffeomorphisms. As a consequence of Moser's theorem, that we have extended to the case of surfaces with boundary in the proof of Proposition 2.3, this class is parametrized by a triple $(p, g, T) \in \mathbb{N}^2 \times \mathbb{R}_+^*$, where $p$ is the number of components of $\partial M$, $g$ the genus of $M$ and $T$ the total area of $M$. Another consequence of this invariance and of Moser's theorem is the symmetry of $Z_M$. Indeed, given any two components of $\partial M$, there exists a diffeomorphism of $M$ that permutes these components, hence an area-preserving diffeomorphism. Thus, for all permutation $\pi \in S_p$, $Z_M(\mathfrak{x}_{\pi(1)}, \ldots, \mathfrak{x}_{\pi(p)}) = Z_M(\mathfrak{x}_1, \ldots, \mathfrak{x}_p)$.

Let us give an expression of the function $Z_M$ that makes clear that it depends on $M$ only through $p$, $g$ and $T$. Consider a graph with only one face on $M$, such that the boundary of this face is $[a_1, b_1] \ldots [a_g, b_g] c_1^{-1} N_1 c_1 \ldots c_p^{-1} N_p c_p$, where $a_i, b_i$ are the edges of a polygonal fundamental domain in the universal cover of a minimal closure of $M$ and each $c_i$ joins $N_i$ to a fixed vertex on the boundary of this fundamental domain. We find

$$Z_{p,g,T}(\mathfrak{x}_1,\ldots,\mathfrak{x}_n) = \tag{5.13}$$

$$= \int_{G^{2g+p}} p_T(y_1^{-1}x_1y_1\ldots y_p^{-1}x_py_p[a_1,b_1]\ldots[a_g,b_g])\, da_1db_1\ldots da_gdb_gdy_1\ldots dy_p$$

where $x_1,\ldots,x_p$ are arbitrary representatives of $\mathfrak{x}_1,\ldots,\mathfrak{x}_p$ and, when $M$ is closed,

$$Z_{0,g,T} = \int_{G^{2g+p}} p_T([a_1,b_1]\ldots[a_g,b_g])\, da_1db_1\ldots da_gdb_g. \tag{5.14}$$

From now on, we will index the function $Z$ by the triple $(p,g,T)$ rather than by the surface $M$.

Formula (5.13) defines $Z_{p,g,T}$ both as a smooth central function on $G^p$ and a continuous function on $(G/\mathrm{Ad})^p$. On the other hand, the symmetry of $Z_{p,g,T}$ is not obvious in this form. Using character expansions, it is possible to give a manifestly symmetric expression of $Z_{p,g,T}$. The reader who is not familiar with the characters of a compact Lie group can read the beginning of Section 4.1.3 before going further. Using the expansion of the heat kernel established in Proposition 4.8, we transform respectively (5.13) and (5.14) into:

$$Z_{p,g,T}(\mathfrak{x}_1,\ldots,\mathfrak{x}_p) = \sum_{\alpha \in \hat{G}} (\dim \alpha)^{2-2g} e^{-\frac{c_\alpha}{2}T} \prod_{i=1}^{p} \frac{\chi_\alpha(\mathfrak{x}_i)}{\dim \alpha}, \tag{5.15}$$

$$Z_{0,g,T} = \sum_{\alpha \in \hat{G}} (\dim \alpha)^{2-2g} e^{-\frac{c_\alpha}{2}T}. \tag{5.16}$$

Before putting these results into a theorem, let us recall that $G/\mathrm{Ad}$ is endowed with the image measure of the Haar measure by the canonical projection $G \longrightarrow G/\mathrm{Ad}$. Most of this theorem has already been stated by E. Witten in [44].

THEOREM 5.27. *For each $(p,g,T) \in \mathbb{N}^2 \times \mathbb{R}_+^*$, the function $Z_{g,p,T}$ is a continuous symmetric function on $(G/\mathrm{Ad})^p$ and its lift to $G^p$ is smooth. Moreover, for any $(p',g',T')$ and any $\mathfrak{x}_1,\ldots,\mathfrak{x}_p, \mathfrak{x}_1',\ldots,\mathfrak{x}_{p'}'$ in $G/\mathrm{Ad}$, the following relations hold:*

$$\int_{G/\mathrm{Ad}} Z_{p+1,g,T}(\mathfrak{x}_1,\ldots,\mathfrak{x}_p,\mathfrak{x}) Z_{p'+1,g',T'}(\mathfrak{x}^{-1},\mathfrak{x}_1',\ldots,\mathfrak{x}_{p'}')\, d\mathfrak{x} =$$

$$Z_{p+p',g+g',T+T'}(\mathfrak{x}_1,\ldots,\mathfrak{x}_p,\mathfrak{x}_1',\ldots,\mathfrak{x}_{p'}'), \tag{5.17}$$

$$\int_{G/\mathrm{Ad}} Z_{p+2,g,T}(\mathfrak{x}_1,\ldots,\mathfrak{x}_p,\mathfrak{x},\mathfrak{x}^{-1})\, d\mathfrak{x} = Z_{p,g+1,T}(\mathfrak{x}_1,\ldots,\mathfrak{x}_p). \tag{5.18}$$

PROOF. The symmetry and continuity of $Z_{p,g,T}$ have already been discussed. The relation (5.17) is a consequence of Proposition 5.3. Indeed, in this proposition, the number of components of the boundary of $M$ is $p_1 + p_2 - 2$, where $p_1$ and $p_2$ are the number of components of $\partial M_1$ and $\partial M_2$ and its genus and total area are the sums of those of $M_1$ and $M_2$. This gives:

$$\int_{G/\mathrm{Ad}} Z_{p+1,g,T}(\mathfrak{x}_1,\ldots,\mathfrak{x}_p,\mathfrak{x}) Z_{p'+1,g',T'}(\mathfrak{x}^{-1},\mathfrak{x}_1',\ldots,\mathfrak{x}_{p'}')\, d\mathfrak{x} =$$

$$\int_{G/\mathrm{Ad}} Z_M(\mathfrak{x}_1,\ldots,\mathfrak{x}_p,\mathfrak{x},\mathfrak{x}_1',\ldots,\mathfrak{x}_{p'}')\, d\mathfrak{x}.$$

In this last partition function, the variable $\mathfrak{x}$ corresponds to an interior loop of $M$, not to a component of the boundary. If we compute this function using the discrete theory, we see that the condition with respect to this interior loop disappears if we integrate over $\mathfrak{x}$, so that the last integral is exactly equal to $Z_{p+p',g+g',T+T'}(\mathfrak{x}_1,\ldots,\mathfrak{x}_p,\mathfrak{x}'_1,\ldots,\mathfrak{x}'_{p'})$.

Similarly, the relation (5.18) is a consequence of Proposition 5.16. In this gluing operation, the surface $M_1$ looses two components of its boundary and gains one handle. Thus,

$$\int_{G/\mathrm{Ad}} Z_{p+2,g,T}(\mathfrak{x}_1,\ldots,\mathfrak{x}_p,\mathfrak{x},\mathfrak{x}^{-1})\,d\mathfrak{x} = \int_{G/\mathrm{Ad}} Z_M(\mathfrak{x}_1,\ldots,\mathfrak{x}_p,\mathfrak{x})\,d\mathfrak{x}.$$

Just as above, $\mathfrak{x}$ disappears when we integrate against $d\mathfrak{x}$ because it corresponds to an interior loop and we find $Z_{p,g+1,T}(\mathfrak{x}_1,\ldots,\mathfrak{x}_p)$. □

**5.4.2. Building bricks of the theory.** Relations (5.17) and (5.18) are the analytical counterparts of the behaviour of the Yang-Mills measure under two basic surgery operations that we have considered earlier in this chapter. It is well-known that very few elementary surfaces are required in order to be able to build any surface by a finite sequence of these basic operations, for example a disk and a three-holed sphere (see Fig. 2). It is not surprising that a corresponding result holds for the conditional partition functions.

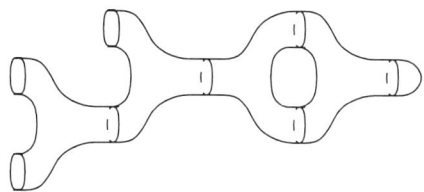

FIGURE 2. An example of decomposition in three-holed spheres and disks.

PROPOSITION 5.28. *The family of functions $(Z_{p,g,T}, p,g \geq 0, T > 0)$ is completely determined by the functions $(Z_{1,0,T}, T > 0)$ and $(Z_{3,0,T}, T > 0)$ and the relations (5.17) and (5.18).*

PROOF. We choose $g,p$ and construct the functions $Z_{g,p,T}$, $T > 0$, starting with the functions $Z_{1,0,T}$ and $Z_{3,0,T}$.

Suppose first that $p + 2g \geq 3$. In this case, repeated applications of (5.17) to the function $Z_{3,0,T}$ allow to compute $Z_{p+2g,0,T}$ for any $T > 0$. Now, $g$ applications of (5.18) to $Z_{p+2g,0,T}$ give the function $Z_{p,g,T}$.

The case $p + 2g = 2$ happens when $(p,g) = (0,1)$ or $(2,0)$. The first case is that of a closed torus. Start with a three-holed sphere and glue two components of its boundary. We get a torus with one hole. This corresponds to (5.18) applied to $Z_{3,0,T}$ to get $Z_{1,1,T}$. Now, it remains to glue a disk on the hole of the torus. In other words, the relation (5.17) applied to $Z_{1,1,T}$ and $Z_{1,0,T}$ gives $Z_{0,1,T}$. The case $(p,g) = (2,0)$ is that of a cylinder, which is obtained by gluing a disk on a three-holed sphere. So, $Z_{2,0,T}$ is obtained by applying (5.17) to $Z_{3,0,T}$ and $Z_{1,0,T}$.

The case $p + 2g = 1$ happens only when $(p, g) = (1, 0)$ and the corresponding function is one of our building bricks.

Finally, $p = g = 0$ is a closed sphere, which can be obtained by gluing two disks together. So, (5.17) applied to $Z_{1,0,T}$ gives $Z_{0,0,T}$. □

The natural question arising from this result is to identify the elementary functions $(Z_{1,0,T}, T > 0)$ and $(Z_{3,0,T}, T > 0)$. Let us get rid of the time dependence by showing that $Z_{p,g,T}$ is a solution of the heat equation in each of its variables.

PROPOSITION 5.29. *For all $(p, g) \in \mathbb{N}^2$, all $T, T' > 0$ and all $x_1, \ldots, x_{p-1}, x$, the following relation holds:*

$$e^{T'\frac{\Delta}{2}} Z_{p,g,T'}(x_1, \ldots, x_{p-1}, \cdot)(x) = Z_{p,g,T+T'}(x_1, \ldots, x_{p-1}, x),$$

*where $\Delta$ is the bi-invariant Laplace operator on $G$.*

PROOF. If we remember that $\Delta \chi_\alpha = -c_\alpha \chi_\alpha$ for all irreducible representation $\alpha$, this assertion is a simple consequence of (5.15). This can also be seen directly on the relation (5.13). □

This lemma shows that the meaning of $Z_{p,g,T}$, if there is one, is contained in the formal limit $\lim_{T \to 0} Z_{p,g,T}$. For example, $Z_{1,0,0}$ ought to be the density with respect to the Haar measure of the distribution of the random holonomy along the boundary of a disk of area zero, that is, the Dirac mass at the unit element.

The interpretation of $Z_{3,0,0}$ is less obvious and more interesting. Let us look at a three-holed sphere with a very small area (see Fig. 3). At the $T \to 0$ limit, there remains only a graph and it appears that each component of the boundary is equivalent to the product of the two others. This suggests the existence of a close link between $Z_{3,0,0}$ and the multiplication in $G$. Of course, there is no multiplication on $G/\mathrm{Ad}$, however, the convolution product of two central functions $f, g \in L^2(G, dx)$ defined by $f * g(x) = \int_G f(y) g(y^{-1} x) \, dy$ is still a central function. Let us denote by $L^2(G)^G$ the space of square-integrable central functions on $G$.

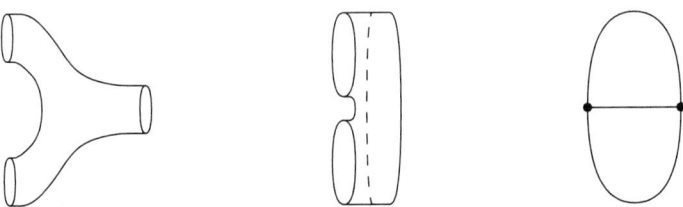

FIGURE 3. A thinner and thinner three-holed sphere.

PROPOSITION 5.30. *1. For all $T > 0$, the function $Z_{1,0,T}$ is the projection on $G/\mathrm{Ad}$ of the heat kernel $p_T$ on $G$: if $f$ is a function of $L^2(G)^G$, then*

$$\int_G f(x) Z_{1,0,T}(x, \cdot) \, dx = e^{T\frac{\Delta}{2}} f.$$

*2. Let $f_1$ and $f_2$ be two functions of $L^2(G)^G$. Then*

$$\int_{G^2} f_1(x_1) f_2(x_2) Z_{3,0,T}(x_1, x_2, \cdot) \, dx_1 dx_2 = e^{T\frac{\Delta}{2}} (f_1 * f_2).$$

This result provides us with a nice interpretation: formally, $Z_{1,0,0}$ is the Dirac mass at 1 and $Z_{3,0,0}$ is the distributional kernel of the convolution operator $*$ : $L^2(G)^G \otimes L^2(G)^G \longrightarrow L^2(G)^G$. From this point of view, the commutativity of the convolution product finds its geometric counterpart in the fact that two holes of a three-holed sphere are indistinguishable under area-preserving diffeomorphisms.

PROOF. 1. Any expression of $Z_{1,0,T}$, for example (5.13), proves this assertion.
2. We use the fact that any central square-integrable function can be expanded as a series of characters. Thus, it is sufficient to prove the theorem when $f_1$ and $f_2$ are the characters of two irreducible representations $\alpha$ and $\beta$. We use the expansion of $Z_{3,0,T}$ given by (5.15).

$$\int_{G^2} \chi_\alpha(x_1)\chi_\beta(x_2) Z_{3,0,T}(x_1, x_2, x)\, dx_1 dx_2 =$$

$$= \sum_{\gamma \in \widehat{G}} (\dim \gamma)^{-1} e^{-\frac{c_\gamma}{2}T} \chi_\gamma(x) \int_G \chi_\alpha(x_1)\chi_\gamma(x_1)\, dx_1 \int_G \chi_\beta(x_2)\chi_\gamma(x_2)\, dx_2$$

$$= \sum_{\gamma \in \widehat{G}} (\dim \gamma)^{-1} e^{-\frac{c_\gamma}{2}T} \chi_\gamma(x) \delta_{\alpha,\gamma} \delta_{\beta,\gamma}$$

$$= \delta_{\alpha,\beta} e^{-\frac{c_\alpha}{2}T} \frac{\chi_\alpha(x)}{\dim \alpha}. \tag{5.19}$$

On the other hand, the orthogonality relations between characters imply:

$$\chi_\alpha * \chi_\beta(x) = \int_G \chi_\alpha(y)\chi_\beta(y^{-1}x)\, dy = \delta_{\alpha,\beta} \frac{\chi_\alpha(x)}{\dim \alpha}.$$

Finally, the fact that $\Delta \chi_\alpha = -c_\alpha \chi_\alpha$ shows that the expression (5.19) is exactly equal to $\exp(T\frac{\Delta}{2})(\chi_\alpha * \chi_\beta)$. □

# Bibliography

[1] ADAMS, C. R., AND LEWY, H. On convergence in length. *Duke Math. J. 1* (1935), 19–26.
[2] ALBEVERIO, S., HØEGH-KROHN, R., AND HOLDEN, H. Stochastic Lie group-valued measures and their relations to stochastic curve integrals, gauge fields and Markov cosurfaces. In *Stochastic processes—mathematics and physics (Bielefeld, 1984)*. Springer, Berlin, 1986, pp. 1–24.
[3] ALBEVERIO, S., HØEGH-KROHN, R., AND HOLDEN, H. Stochastic multiplicative measures, generalized Markov semigroups, and group-valued stochastic processes and fields. *J. Funct. Anal. 78*, 1 (1988), 154–184.
[4] ASHTEKAR, A., AND ISHAM, C. J. Representations of the holonomy algebras of gravity and nonabelian gauge theories. *Classical Quantum Gravity 9*, 6 (1992), 1433–1467.
[5] ASHTEKAR, A., AND LEWANDOWSKI, J. Representation theory of analytic holonomy $C^*$-algebras. In *Knots and quantum gravity (Riverside, CA, 1993)*. Oxford Univ. Press, New York, 1994, pp. 21–61.
[6] ASHTEKAR, A., AND LEWANDOWSKI, J. Projective techniques and functional integration for gauge theories. *J. Math. Phys. 36*, 5 (1995), 2170–2191.
[7] ASHTEKAR, A., MAROLF, D., AND MOURÃO, J. Integration on the space of connections modulo gauge transformations. In *Proceedings of the Cornelius Lanczos International Centenary Conference (Raleigh, NC, 1993)* (Philadelphia, PA, 1994), SIAM, pp. 143–160.
[8] AUBIN, T. *Some nonlinear problems in Riemannian geometry*. Springer-Verlag, Berlin, 1998.
[9] BAEZ, J. C., AND SAWIN, S. Functional integration on spaces of connections. *J. Funct. Anal. 150*, 1 (1997), 1–26.
[10] BAEZ, J. C., AND SAWIN, S. Diffeomorphism-invariant spin network states. *J. Funct. Anal. 158*, 2 (1998), 253–266.
[11] BECKER, C. Wilson loops in two-dimensional space-time regarded as white noise. *J. Funct. Anal. 134*, 2 (1995), 321–349.
[12] BECKER, C., AND SENGUPTA, A. Sewing Yang-Mills measures and moduli spaces over compact surfaces. *J. Funct. Anal. 152*, 1 (1998), 74–99.
[13] BLEECKER, D. *Gauge theory and variational principles*. Addison-Wesley Publishing Co., Reading, Mass., 1981.
[14] BOTT, R., AND TU, L. W. *Differential forms in algebraic topology*. Springer-Verlag, New York, 1982.
[15] BOURBAKI, N. *Éléments de mathématique. Théorie des ensembles*. Hermann, Paris, 1970.
[16] BRÖCKER, T., AND TOM DIECK, T. *Representations of compact Lie groups*. Springer-Verlag, New York, 1995. Translated from the German manuscript, Corrected reprint of the 1985 translation.
[17] DELLACHERIE, C. *Analytic sets*. Academic Press Inc. [Harcourt Brace Jovanovich Publishers], London, 1980. Lectures delivered at a Conference held at University College, University of London, London, July 16–29, 1978.
[18] DELLACHERIE, C., AND MEYER, P.-A. *Probabilités et potentiel*. Hermann, Paris, 1975. Chapitres I à IV, Édition entièrement refondue, Publications de l'Institut de Mathématique de l'Université de Strasbourg, No. XV, Actualités Scientifiques et Industrielles, No. 1372.
[19] DO CARMO, M. P. A. *Riemannian geometry*. Birkhäuser Boston Inc., Boston, MA, 1992. Translated from the second Portuguese edition by Francis Flaherty.
[20] DRIVER, B. K. $YM_2$: continuum expectations, lattice convergence, and lassos. *Comm. Math. Phys. 123*, 4 (1989), 575–616.

[21] DRIVER, B. K. Two-dimensional Euclidean quantized Yang-Mills fields. In *Probability models in mathematical physics (Colorado Springs, CO, 1990)*. World Sci. Publishing, Teaneck, NJ, 1991, pp. 21–36.

[22] FINE, D. S. Quantum Yang-Mills on the two-sphere. *Comm. Math. Phys. 134*, 2 (1990), 273–292.

[23] FINE, D. S. Quantum Yang-Mills on a Riemann surface. *Comm. Math. Phys. 140*, 2 (1991), 321–338.

[24] FLEISCHHACK, C. On the support of physical measures in gauge theories.

[25] GRAY, A. *Tubes*. Addison-Wesley Publishing Company Advanced Book Program, Redwood City, CA, 1990.

[26] GROSS, L. A Poincaré lemma for connection forms. *J. Funct. Anal. 63*, 1 (1985), 1–46.

[27] GROSS, L. The Maxwell equations for Yang-Mills theory. In *Mathematical quantum field theory and related topics (Montreal, PQ, 1987)*. Amer. Math. Soc., Providence, RI, 1988, pp. 193–203.

[28] GROSS, L., KING, C., AND SENGUPTA, A. Two-dimensional Yang-Mills theory via stochastic differential equations. *Ann. Physics 194*, 1 (1989), 65–112.

[29] IKEDA, N., AND WATANABE, S. *Stochastic differential equations and diffusion processes*, second ed. North-Holland Publishing Co., Amsterdam, 1989.

[30] KOBAYASHI, S., AND NOMIZU, K. *Foundations of differential geometry. Vol. I*. John Wiley & Sons Inc., New York, 1996. Reprint of the 1963 original, A Wiley-Interscience Publication.

[31] LÉVY, T. Construction et étude à l'échelle microscopique de la mesure de Yang-Mills sur les surfaces compactes. *C. R. Acad. Sci. Paris Sér. I Math. 330*, 11 (2000), 1019–1024.

[32] MIGDAL, A. A. Recursion equations in gauge field theories. *Sov. Phys. JETP 42*, 3 (1975), 413–418.

[33] MOSER, J. On the volume elements on a manifold. *Trans. Amer. Math. Soc. 120* (1965), 286–294.

[34] RADÓ, T. *Length and Area*. American Mathematical Society, New York, 1948.

[35] RAO, M. M. Projective limits of probability spaces. *J. Multivariate Anal. 1*, 1 (1971), 28–57.

[36] SENGUPTA, A. The Yang-Mills measure for $S^2$. *J. Funct. Anal. 108*, 2 (1992), 231–273.

[37] SENGUPTA, A. Gauge invariant functions of connections. *Proc. Amer. Math. Soc. 121*, 3 (1994), 897–905.

[38] SENGUPTA, A. Gauge theory on compact surfaces. *Mem. Amer. Math. Soc. 126*, 600 (1997), viii+85.

[39] SENGUPTA, A. Yang-Mills on surfaces with boundary: quantum theory and symplectic limit. *Comm. Math. Phys. 183*, 3 (1997), 661–705.

[40] SIMON, B. *Representations of finite and compact groups*. American Mathematical Society, Providence, RI, 1996.

[41] SPANIER, E. H. *Algebraic topology*. Springer-Verlag, New York, 1981. Corrected reprint.

[42] TAYLOR, M. E. *Partial differential equations. I*. Springer-Verlag, New York, 1996. Basic theory.

[43] VAROPOULOS, N. T., SALOFF-COSTE, L., AND COULHON, T. *Analysis and geometry on groups*. Cambridge University Press, Cambridge, 1992.

[44] WITTEN, E. On quantum gauge theories in two dimensions. *Comm. Math. Phys. 141*, 1 (1991), 153–209.

[45] YANG, C. N., AND MILLS, R. L. Conservation of isotopic spin and isotopic gauge invariance. *Physical Rev. (2) 96* (1954), 191–195.

## Editorial Information

To be published in the *Memoirs*, a paper must be correct, new, nontrivial, and significant. Further, it must be well written and of interest to a substantial number of mathematicians. Piecemeal results, such as an inconclusive step toward an unproved major theorem or a minor variation on a known result, are in general not acceptable for publication. Papers appearing in *Memoirs* are generally longer than those appearing in *Transactions*, which shares the same editorial committee.

As of July 1, 2003, the backlog for this journal was approximately 3 volumes. This estimate is the result of dividing the number of manuscripts for this journal in the Providence office that have not yet gone to the printer on the above date by the average number of monographs per volume over the previous twelve months, reduced by the number of volumes published in four months (the time necessary for preparing a volume for the printer). (There are 6 volumes per year, each containing at least 4 numbers.)

A Consent to Publish and Copyright Agreement is required before a paper will be published in the *Memoirs*. After a paper is accepted for publication, the Providence office will send a Consent to Publish and Copyright Agreement to all authors of the paper. By submitting a paper to the *Memoirs*, authors certify that the results have not been submitted to nor are they under consideration for publication by another journal, conference proceedings, or similar publication.

## Information for Authors

*Memoirs* are printed from camera copy fully prepared by the author. This means that the finished book will look exactly like the copy submitted.

The paper must contain a *descriptive title* and an *abstract* that summarizes the article in language suitable for workers in the general field (algebra, analysis, etc.). The *descriptive title* should be short, but informative; useless or vague phrases such as "some remarks about" or "concerning" should be avoided. The *abstract* should be at least one complete sentence, and at most 300 words. Included with the footnotes to the paper should be the 2000 *Mathematics Subject Classification* representing the primary and secondary subjects of the article. The classifications are accessible from **www.ams.org/msc/**. The list of classifications is also available in print starting with the 1999 annual index of *Mathematical Reviews*. The Mathematics Subject Classification footnote may be followed by a list of *key words and phrases* describing the subject matter of the article and taken from it. Journal abbreviations used in bibliographies are listed in the latest *Mathematical Reviews* annual index. The series abbreviations are also accessible from **www.ams.org/publications/**. To help in preparing and verifying references, the AMS offers MR Lookup, a Reference Tool for Linking, at **www.ams.org/mrlookup/**. When the manuscript is submitted, authors should supply the editor with electronic addresses if available. These will be printed after the postal address at the end of the article.

**Electronically prepared manuscripts.** The AMS encourages electronically prepared manuscripts, with a strong preference for $\mathcal{AMS}$-LaTeX. To this end, the Society has prepared $\mathcal{AMS}$-LaTeX author packages for each AMS publication. Author packages include instructions for preparing electronic manuscripts, the *AMS Author Handbook*, samples, and a style file that generates the particular design specifications of that publication series. Though $\mathcal{AMS}$-LaTeX is the highly preferred format of TeX, author packages are also available in $\mathcal{AMS}$-TeX.

Authors may retrieve an author package from e-MATH starting from **www.ams.org/tex/** or via FTP to **ftp.ams.org** (login as **anonymous**, enter username as password, and type **cd pub/author-info**). The *AMS Author Handbook* and the *Instruction Manual* are available in PDF format following the author packages link from **www.ams.org/tex/**. The author package can be obtained free of charge by sending email to **pub@ams.org** (Internet) or from the Publication Division, American Mathematical Society, 201 Charles St., Providence, RI 02904, USA. When requesting an author package, please specify $\mathcal{AMS}$-LaTeX or $\mathcal{AMS}$-TeX, Macintosh or IBM (3.5) format, and the publication in which your paper will appear. Please be sure to include your complete mailing address.

**Sending electronic files.** After acceptance, the source file(s) should be sent to the Providence office (this includes any TeX source file, any graphics files, and the DVI or PostScript file).

Before sending the source file, be sure you have proofread your paper carefully. The files you send must be the EXACT files used to generate the proof copy that was accepted for publication. For all publications, authors are required to send a printed copy of their paper, which exactly matches the copy approved for publication, along with any graphics that will appear in the paper.

TeX files may be submitted by email, FTP, or on diskette. The DVI file(s) and PostScript files should be submitted only by FTP or on diskette unless they are encoded properly to submit through email. (DVI files are binary and PostScript files tend to be very large.)

Electronically prepared manuscripts can be sent via email to **pub-submit@ams.org** (Internet). The subject line of the message should include the publication code to identify it as a Memoir. TeX source files, DVI files, and PostScript files can be transferred over the Internet by FTP to the Internet node **e-math.ams.org** (130.44.1.100).

**Electronic graphics.** Comprehensive instructions on preparing graphics are available at **www.ams.org/jourhtml/graphics.html**. A few of the major requirements are given here.

Submit files for graphics as EPS (Encapsulated PostScript) files. This includes graphics originated via a graphics application as well as scanned photographs or other computer-generated images. If this is not possible, TIFF files are acceptable as long as they can be opened in Adobe Photoshop or Illustrator. No matter what method was used to produce the graphic, it is necessary to provide a paper copy to the AMS.

Authors using graphics packages for the creation of electronic art should also avoid the use of any lines thinner than 0.5 points in width. Many graphics packages allow the user to specify a "hairline" for a very thin line. Hairlines often look acceptable when proofed on a typical laser printer. However, when produced on a high-resolution laser imagesetter, hairlines become nearly invisible and will be lost entirely in the final printing process.

Screens should be set to values between 15% and 85%. Screens which fall outside of this range are too light or too dark to print correctly. Variations of screens within a graphic should be no less than 10%.

**Inquiries.** Any inquiries concerning a paper that has been accepted for publication should be sent directly to the Electronic Prepress Department, American Mathematical Society, 201 Charles St., Providence, RI 02904, USA.

# Editors

This journal is designed particularly for long research papers, normally at least 80 pages in length, and groups of cognate papers in pure and applied mathematics. Papers intended for publication in the *Memoirs* should be addressed to one of the following editors. In principle the Memoirs welcomes electronic submissions, and some of the editors, those whose names appear below with an asterisk (*), have indicated that they prefer them. However, editors reserve the right to request hard copies after papers have been submitted electronically. Authors are advised to make preliminary email inquiries to editors about whether they are likely to be able to handle submissions in a particular electronic form.

***Algebra** to ROBERT GURALNICK, Department of Mathematics, University of Southern California, Los Angeles, CA 90089-1113; email: `guralnic@math.usc.edu`

**Algebraic geometry** to DAN ABRAMOVICH, Department of Mathematics, Boston University, 111 Cummington St., Boston, MA 02215; email: `abramovic@bu.edu`

**Algebraic topology and cohomology of groups** to STEWART PRIDDY, Department of Mathematics, Northwestern University, 2033 Sheridan Road, Evanston, IL 60208-2730; email: `priddy@math.nwu.edu`

**Combinatorics and Lie theory** to SERGEY FOMIN, Department of Mathematics, University of Michigan, Ann Arbor, Michigan 48109-1109; email: `fomin@umich.edu`

**Complex analysis and complex geometry** to DUONG H. PHONG, Department of Mathematics, Columbia University, 2990 Broadway, New York, NY 10027-0029; email: `phong@math.columbia.edu`

***Differential geometry and global analysis** to LISA C. JEFFREY, Department of Mathematics, University of Toronto, 100 St. George St., Toronto, ON Canada M5S 3G3; email: `jeffrey@math.toronto.edu`

**Dynamical systems and ergodic theory** to ROBERT F. WILLIAMS, Department of Mathematics, University of Texas, Austin, Texas 78712-1082; email: `bob@math.utexas.edu`

***Geometric analysis** to TOBIAS COLDING, Courant Institute, New York University, 251 Mercer St., New York, NY 10012; email: `colding@cims.nyu.edu`

**Harmonic analysis** to ALEXANDER NAGEL, Department of Mathematics, University of Wisconsin, 480 Lincoln Drive, Madison, WI 53706-1313; email: `nagel@math.wisc.edu`

**Harmonic analysis, representation theory, and Lie theory** to ROBERT J. STANTON, Department of Mathematics, The Ohio State University, 231 West 18th Avenue, Columbus, OH 43210-1174; email: `stanton@math.ohio-state.edu`

***Logic** to STEFFEN LEMPP, Department of Mathematics, University of Wisconsin, 480 Lincoln Drive, Madison, Wisconsin 53706-1388; email: `lempp@math.wisc.edu`

**Number theory** to HAROLD G. DIAMOND, Department of Mathematics, University of Illinois, 1409 W. Green St., Urbana, IL 61801-2917; email: `diamond@math.uiuc.edu`

***Ordinary differential equations, and applied mathematics** to PETER W. BATES, Department of Mathematics, Michigan State University, East Lansing, MI 48824-1027; email: `peter@math.msu.edu`

***Partial differential equations** to PATRICIA E. BAUMAN, Department of Mathematics, Purdue University, West Lafayette, IN 47907-1395' email: `bauman@math.purdue.edu`

***Probability and statistics** to KRZYSZTOF BURDZY, Department of Mathematics, University of Washington, Box 354350, Seattle, Washington 98195-4350; email: `burdzy@math.washington.edu`

***Real analysis and partial differential equations** to DANIEL TATARU, Department of Mathematics, University of California, Berkeley, Berkeley, CA 94720; email: `tataru@math.berkeley.edu`

**All other communications to the editors** should be addressed to the Managing Editor, WILLIAM BECKNER, Department of Mathematics, University of Texas, Austin, TX 78712-1082; email: `beckner@math.utexas.edu`.

# Titles in This Series

790 **Thierry Lévy,** Yang-Mills measure on compact surfaces, 2003

789 **Helge Glöckner,** Positive definite functions on infinite-dimensional convex cones, 2003

788 **Robert Denk, Matthias Hieber, and Jan Prüss,** $\mathcal{R}$-boundedness, Fourier multipliers and problems of elliptic and parabolic type, 2003

787 **Michael Cwikel, Per G. Nilsson, and Gideon Schechtman,** Interpolation of weighted Banach lattices/A characterization of relatively decomposable Banach lattices, 2003

786 **Arnd Scheel,** Radially symmetric patterns of reaction-diffusion systems, 2003

785 **R. R. Bruner and J. P. C. Greenlees,** The connective K-theory of finite groups, 2003

784 **Desmond Sheiham,** Invariants of boundary link cobordism, 2003

783 **Ethan Akin, Mike Hurley, and Judy A. Kennedy,** Dynamics of topologically generic homeomorphisms, 2003

782 **Masaaki Furusawa and Joseph A. Shalika,** On central critical values of the degree four $L$-functions for GSp(4): The Fundamental Lemma, 2003

781 **Marcin Bownik,** Anisotropic Hardy spaces and wavelets, 2003

780 **S. Marmi and D. Sauzin,** Quasianalytic monogenic solutions of a cohomological equation, 2003

779 **Hansjörg Geiges,** $h$-principles and flexibility in geometry, 2003

778 **David B. Massey,** Numerical control over complex analytic singularities, 2003

777 **Robert Lauter,** Pseudodifferential analysis on conformally compact spaces, 2003

776 **U. Haagerup, H. P. Rosenthal, and F. A. Sukochev,** Banach embedding properties of non-commutative $L^p$-spaces, 2003

775 **P. Lochak, J.-P. Marco, and D. Sauzin,** On the splitting of invariant manifolds in multidimensional near-integrable Hamiltonian systems, 2003

774 **Kai A. Behrend,** Derived $\ell$-adic categories for algebraic stacks, 2003

773 **Robert M. Guralnick, Peter Müller, and Jan Saxl,** The rational function analogue of a question of Schur and exceptionality of permutation representations, 2003

772 **Katrina Barron,** The moduli space of $N=1$ superspheres with tubes and the sewing operation, 2003

771 **Shigenori Matsumoto,** Affine flows on 3-manifolds, 2003

770 **W. N. Everitt and L. Markus,** Elliptic partial differential operators and symplectic algebra, 2003

769 **Jie Wu,** Homotopy theory of the suspensions of the projective plane, 2003

768 **R. Höpfner and E. Löcherbach,** Limit theorems for null recurrent Markov processes, 2003

767 **Po Hu,** $S$-modules in the category of schemes, 2003

766 **Su Gao and Alexander S. Kechris,** On the classification of Polish metric spaces up to isometry, 2003

765 **Robert Bieri and Ross Geoghegan,** Connectivity properties of group actions on non-positively curved spaces, 2003

764 **J. Spandaw,** Noether-Lefschetz problems for degeneracy loci, 2003

763 **Yasuyuki Kachi and Eiichi Sato,** Segre's reflexivity and an inductive characterization os hyperquadrics, 2002

762 **Leiba Rodman, Ilya M. Spitkovsky, and Hugo Woerdeman,** Abstract band method via factorization, positive and band extensions of multivariable almost periodic matrix functions, and spectral estimation, 2002

761 **Oliver Druet and Emmanuel Hebey,** The $AB$ program in geometric analysis : Sharp Sobolev inequalities and related problems, 2002

## TITLES IN THIS SERIES

760 **Markus Banagl,** Extending intersection homology type invariants to non-Witt spaces, 2002

759 **Donald M. Davis,** From representation theory to homotopy groups, 2002

758 **Alan Forrest, John Hunton, and Johannes Kellendonk,** Topological invariants for projection method patterns, 2002

757 **Douglas Bowman,** $q$-difference operators, orthogonal polynomials, and symmetric expansions, 2002

756 **José Ignacio Cogolludo-Agustín,** Topological invariants of the complement to arrangements of rational plane curves, 2002

755 **M. A. Mandell and J. P. May,** Equivariant orthogonal spectra and $S$-modules, 2002

754 **Edward L. Green, Idun Reiten, and Øyvind Solberg,** Dualities on generalized Koszul algebras, 2002

753 **Daniel Panazzolo,** Desingularization of nilpotent singularities in families of planar vector fields, 2002

752 **Linus Kramer,** Homogeneous spaces, Tits buildings, and isoparametric hypersurfaces, 2002

751 **Bruce Allison, Georgia Benkart, and Yun Gao,** Lie algebras graded by the root systems $BC_r$, $r \geq 2$, 2002

750 **Masaki Izumi and Hideki Kosaki,** Kac algebras arising from composition of subfactors: General theory and classification, 2002

749 **Nanhua Xi,** The based ring of two-sided cells of affine Weyl groups of type $\widetilde{A}_{n-1}$, 2002

748 **Jürgen Ritter and Alfred Weiss,** The lifted root number conjecture and Iwasawa theory, 2002

747 **Armand Borel, Robert Friedman, and John W. Morgan,** Almost commuting elements in compact Lie groups, 2002

746 **Peter Niemann,** Some generalized Kac-Moody algebras with known root multiplicities, 2002

745 **Mikhail A. Lifshits and Werner Linde,** Approximation and entropy numbers of Volterra operators with application to Brownian motion, 2002

744 **Roger Chalkley,** Basic global relative invariants for homogeneous linear differential equations, 2002

743 **Heng Sun,** Spectral decomposition of a covering of $GL(r)$: the Borel case, 2002

742 **J. E. Gilbert, Y. S. Han, J. A. Hogan, J. D. Lakey, D. Weiland, and G. Weiss,** Smooth molecular functions and singular integral operators, 2002

741 **Francisco Santos,** Triangulations of oriented matroids, 2002

740 **Rick Durrett,** Mutual invadability implies coexistence in spatial models, 2002

739 **Georgios K. Alexopoulos,** Sub-Laplacians with drift on Lie groups of polynomial volume growth, 2002

738 **Yasuro Gon,** Generalized Whittaker functions on $SU(2,2)$ with respect to the Siegel parabolic subgroup, 2002

737 **Arjen Doelman, Robert A. Gardner, and Tasso J. Kaper,** A stability index analysis of 1-D patterns of the Gray-Scott model, 2002

736 **Wojciech Chachólski and Jérôme Scherer,** Homotopy theory of diagrams, 2002

735 **Martina Brück, Xi Du, Joonsang Park, and Chuu-Lian Terng,** The submanifold geometries associated to Grassmannian systems, 2002

For a complete list of titles in this series, visit the
AMS Bookstore at **www.ams.org/bookstore/**.